Medication Administration Aide Handbook For Non-Licensed Personnel

Written by: Robert Palmer, RN, BSN

Copyright © 2013

Course Outline

- Course information and certification (Page 4)
- Glossary of terms (Page 5)
- Rights of individuals supported (Page 8)
- Medication refusals (Page 9)
- Body substance precautions (Page 11)
  - MRSA fact sheet
- Vital signs (Page 15)
  - Temperature, pulse, respirations and blood pressure
- Commonly used abbreviations (Page 25)
- Measurements (Page 27)
- Common psychiatric conditions (Page 28)
  - Depression
  - Paranoid Personality Disorder
  - Anxiety Disorders
  - ADHD
  - Mood Disorders
  - Conduct Disorder
- Diabetes (Page 36)
  - General information
  - Hyperglycemia
  - Hypoglycemia
  - Glucagon kits
  - Diabetic nutrition information
  - Glucometer training
- Seizures (Page 44)
  - General information about seizures
  - Sample seizure protocols
  - Seizure logs
  - Diastat information and steps
  - Vagus Nerve Stimulators
- Drug reference material (Page 53)
- Medication classifications (Page 56)
- Adverse reactions (Page 61)
  - Anaphylaxis
  - Epi-Pens
- Psychotropic medications (Page 68)
- Extrapyramidal Symptoms (EPS) (Page 70)
  - Acute Dyskinesia
  - Akathesia
  - Dystonia
  - Pseudo-Parkinsonism
  - Tardive Dyskinesia

- Medication Administration Records (MAR) (Page 73)
  - General information
  - Documenting medication refusals
  - Documenting if a person vomits a medication
  - Documenting medication errors on a MAR
  - Discontinuing a medication
  - Transcribing new medication orders
  - Transcribing and documenting changes in current orders
  - Transcribing medications administered every other day or monthly
  - Transcribing and documenting pulse requirements before certain medications
  - Setting up medications for home visits
- Administration of Medications (Page 89)
  - General rules
  - Step by step instructions
  - Self administration or partial participation of medication administration
  - Administering:
    - Eye medications
    - Ear medications
    - Nasal medications
    - Inhaled medications
    - Topical medications
    - Vaginal medications
    - Rectal medications
    - Prefilled enema bottles
    - Oxygen via nasal cannula or mask
    - Transdermal patches
  - Special instructions for administering medications
    - Sublingual and buccal medications
    - Measuring liquids
- Nursing delegation (Page 108)
- Tube feedings and medications through G/J tubes (Page 110)
- Specialized diets (Page 117)
  - Thickened liquids
  - Mechanically altered diets
  - Pureed diets
- Medication storage (Page 122)
- Controlled substances (Page 123)
- Medication disposal (Page 125)
- Medication errors (Page 127)
- Methods to prevent medication errors (Page 128)
- Guidelines for observing and reporting (Page 130)
  - Signs and symptoms per body system
  - Emergency conditions requiring emergency room visits
  - Poison control notification recommendations
- Contract between instructor and student (Page 134)

Course Information and Certification

- The course length is 16 hours but may be longer to meet the needs of the class participants or the trainer.
- Class size is limited to 15 students.
- Only a Registered Nurse or Licensed Practical Nurse may teach this course.
- Students must be at least 18 years of age and have a high school diploma or GED.
- Students must attend the entire 16 hours of classroom instruction and participate in discussion and activities.
- The student will need to pass a written 50 question, closed book exam at the end of the course with a score of 80% of higher. If the student fails the exam, the instructor may, at his or her discretion, offer a second test within seven days. If the student is unsuccessful with the second exam, he/she must retake the entire 16 hour course.
- The student must complete, to the satisfaction of the instructor, an actual or simulated medication administration med pass.
- Upon completion of the course, but before administering medications the student must be watched at least twice administering medications to Individuals supported. This does not have to be a licensed nurse, but may be a supervisor or delegated employee. The person watching the actual medication administration needs to have at least one year of experience passing medication and is trusted to follow all of the standards taught within this handbook.
- Once you have completed this course you will obtain a certificate. This certificate is valid for two years. At the end of two years you must complete a four (4) hour refresher course in order to maintain your certificate. If you fail to complete a refresher course before your certificate expires, you are not allowed to pass medications from the date that you are expired until you complete a refresher course.
- If a certificate is expired for more than three months, the student will be required to either retake the entire course or challenge the final exam along with adequately demonstrating a simulated medication pass in front of an instructor. The student must also administer medications at least twice to an individual supported by an experienced staff with at least one year of experience who is trusted to follow all of the standards listed in this handbook.
- If any supervisor within the organization or funding source finds that the person holding a medication administration certificate is not safely performing their duties, they shall immediately suspend that person's right to perform such functions. Suspensions may be temporary, but could lead to revocation, depending on the circumstances.
- Revocation of the medication administration certificate will occur if certified personnel do not demonstrate compliance and/or are not performing their duties in a safe manner according to the standards taught in this handbook.
- This course in no way identifies the non-licensed personnel as a Certified Medication Administration Technician. Successful completion of this course allows you to obtain a certificate of completion which must be approved by the organization you are working for along with any State funding sources such the Department of Mental Health.

Glossary of Terms

Developmental Disability - This is a term used to describe lifelong disabilities attributable to mental and/or physical impairments which are manifested before the age of 18. These individuals have needs in the following areas: capacity for independent living, economic self-sufficiency, learning, mobility, receptive and expressive language, self-care and self-direction.

Group Home – A group home is a private residence that is shared by unrelated people who have common characteristics. The people who live in these homes generally have either developmental disabilities, considered medically fragile, abused or neglected youths, or youths with emotional or behavioral problems. The number of individuals supported varies from state to state, but generally no more than 9 people can share one residence.

ISL – Independent Supported Living. This is a private residence shared by unrelated people and is similar to a Group Home. The main difference between an ISL and Group Home is that ISL's cannot have more than 4 Individuals supported in the home. The individuals supported generally have either developmental disabilities, considered medically fragile, abused or neglected youths, or youths with emotional or behavioral problems.

Medically Fragile – People are considered medically fragile when they have higher medical needs and they cannot take care of themselves. These individuals need medical care that is greater than just passing medications and providing encouragement for Activities of Daily Living. Often times the Individuals supported are in wheelchairs, and they may have additional medical needs such as oxygen administration, indwelling or supra pubic catheters, etc. These Individuals need additional oversight in order to keep them safe and healthy.

Host Family – Host Families are homes where up to three individuals supported live with a family. This type of set-up gives the Individual more continuity of care since it is similar to a traditional family setting where staff is not continuously rotating in and out.

IP – Individualized Plan. Every individual supported will have an IP. The IP gives you the information you need to know in order to support the Individual. It includes information such as what they like and don't like, what areas they need support in, how to provide that support, choking protocols, seizure protocols, and objectives the person is working on. You should always read and become familiar with the IP for the individuals supported before you begin working with them.

Addiction – Emotional or psychological dependence on a drug or medication beyond voluntary control.

ADL's – Activities of daily living such as eating, bathing, etc.

Adverse Reaction – An unexpected or unwanted reaction to a medication. It may be sudden or develop over time.

Allergic Reaction – An immune response to a foreign substance resulting in inflammation and/or organ dysfunction. Symptoms can include redness, rash, hives, itching, swelling, fever and yellowing of the skin.

Anaphylaxis – The most severe form of an allergic reaction. It is generally sudden and often fatal. Symptoms include lowering of the blood pressure, respiratory distress and unresponsiveness.

Antagonism – Two medications working against each other. If taken at the same time you will get less of the effects of one or both of the medications.

Contraindications – An existing condition the person has which is not compatible with a certain medication. Example: If you are allergic to eggs you cannot take the flu shot. The flu shot is made from an egg derivative. Therefore, the flu shot is contraindicated for you.

Controlled Substances – Medications considered to be addictive and regulated by Federal Law. These medications are at high risk for being misappropriated.

Cumulative effect – This is a build-up of a medication in your body.

Dermal – Refers to the skin (Dermatological).

Enteric Coated – A special coating on tablets or capsules that dissolve in the small intestines and not the stomach. These pills cannot be broken or crushed, they must be swallowed whole.

Gait – The manner in which a person is walking. Example: limping or staggering.

HIPPA – The federal privacy act that protect the rights of any patient in the United States to have their medical information maintained in confidentiality and protected from being accessed by anyone not directly providing care or services to that Individual.

Hypertension – High blood pressure. Diagnosed by a physician or advanced practice nurse where a person's blood pressure is consistently above the normal readings considered appropriate for that person.

Jaundice – Yellowing of the whites of the eyes, skin and body fluids.

Lethal Dose – This is the amount of medication required to kill you.

Lethargic – Drowsy or sluggish, difficult to stay awake.

Ophthalmic – Pertaining to the eyes.

Orthostatic Hypotension – A change or drop in blood pressure when rising from a sitting or lying position to a standing position. It may result in a person becoming dizzy, falling or losing consciousness.

Otic – Pertaining to the ear.

Over the Counter (OTC) – Medications which can be purchased without a prescription.

Photosensitivity – Easily sunburned. This is a side effect associated with some medications.

PRN – "As Needed". Medications that are only given if a person needs them. Example: If you have a PRN order for Tylenol for pain or fever the medication would not be given routinely, but only if the person experienced pain or fever.

Synergism – Two medications that work together for a desired effect. Example: calcium and vitamin D. If calcium is taken by itself a small amount of calcium will be absorbed. If vitamin D is taken at the same time a larger amount of the medication is absorbed.

Tolerance – A person may develop a tolerance to a medication if they have taken the same medication for a long period of time. The doctor may prescribe an increased dose in order to get the same effect that was produced previously.

Topical Medications – Medication applied to the skin or mucous membranes (eyes, nose, ears, etc).

Toxic – Poisoning. A concentration of a medication in a person's blood that has become high enough to cause harm.

Rights of Individuals Supported

- Individuals supported have the same rights as any other person. There may be times when rights have to be limited for the individual's safety. If a right has to be limited, your agency will have steps that must occur in order for any right to be limited. In addition, if a right is limited, there must be steps in place so that the Individual supported can work towards regaining this right.

- Individuals supported have the right to look at their medical records. This includes any daily notes written about them, their IP and the medication administration record. While they do have the right to look at their records, they do not have the right to review any other individual's documents.

- All records of the Individual supported must be kept confidential. No one should have access to this information except the people responsible for the daily care of this individual, supervisors of the Individual supported, Auditors, or Law Enforcement as needed due to a specific event.

- Individuals supported have the right to choose their own physicians and be as involved in their treatment plan as possible. They also have the right to have all treatments explained to them in a manner that they can understand.

- If Abuse or Neglect is suspected, you are considered a mandated reporter. If you fail to report suspected abuse or neglect, you are not looking out for the best interest of the person supported. By allowing the suspected abuse or neglect to occur you may be subject to the same disciplinary action as the person who actually did the abuse or neglect (if that person is found to be guilty). Report suspected abuse or neglect to either your direct supervisor, their supervisor, a case manager, or the Division of Aging for your State. If you are working with children you may contact your State's Children's Division. The agency you are working with is not allowed to retaliate against you for reporting suspected abuse or neglect. The Division of Aging and Children's Division will allow anonymous reporting.

Medication Refusals

Every medication prescribed for the individual supported is important for their general health. Caregivers need to make every attempt possible to insure that the individuals receive the medications they are prescribed. With this being said, <u>Individuals Supported have the right to refuse their medications</u>. Before you mark a medication down as a refusal, you should make several attempts to get the person to take their medications.

- ➤ Ask the reason they don't want to take the medication. You may be interrupting an activity they are enjoying and they may not want to stop long enough to take the medication. Try offering the medication again during a break or a commercial (if they are watching television). If they feel the medication is upsetting their stomach, offer the medication with milk or crackers as long as these are not contraindicated with the medication being given.
- ➤ Educate the Individual on why the medication was prescribed and possible complications that could occur if they don't take the medication.
    - ○ "This medication is to help prevent you from having seizures. Seizures are really hard on your body and you know how you feel after you have a seizure."
- ➤ Offer the medication a few times throughout your "window of opportunity". Most states allow a one hour window before and after the scheduled medication time. If a medication is scheduled for 8:00 AM, you can administer the medication anytime from 7:00 AM to 9:00 AM. The Individual may refuse the medication at 7:00 AM, but by 9:00 he/she may accept it if you offer it again.
- ➤ If you are working with another staff person, you may have that person try and offer the medications a short time later.
    - ○ "Hey, John, I saw you haven't taken your noon medications yet, I have them for you."
- ➤ Seek advice from your supervisor, co-workers, or community nurse on techniques that may be offered for this specific Individual. It may include giving the medication with a desired drink or food or following a specific routine.
- ➤ As a last resort the medication will be documented as a refusal.

Any time a medication is refused the supervisor, nurse and doctor needs to be aware. Always follow your agencies' policy and procedures for reporting medication refusals. Education regarding the medication needs to occur at every level starting with the Direct Support staff then moving to the supervisor, nurse, case manager, and doctor.

<u>You cannot force an Individual supported to take their medications and you cannot hide medications in their food.</u> A refusal is a refusal and if you hide their medication in their food you are violating their rights. If a medication must be given for the health and safety of the Individual supported, the individual will need to be placed in an environment where medications can be forced. This cannot be done in an ISL or Group Home setting.

Medication refusals are <u>NOT</u> medication errors. The refusal will be documented on the Medication Administration Record. Any time an Individual refuses their medication the daily notes need to include all attempts that were used prior to the medication(s) being listed as refusal.

Practice Session

1. The doctor prescribes a higher dose of a medication a person has been taking for a long time because the individual reports he is not getting same effect from the medication that he used to. This is an example of:

    a. Synergism
    b. Tolerance
    c. Cumulative Effect
    d. Antagonism

2. Upon successful completion of this medication course, the certificate issued is valid for ____ years before a four hour refresher course is required.

    a. 1 year
    b. 2 years
    c. 3 years
    d. 4 years

3. This is the most severe form of a drug reaction, it is often sudden and fatal.

    a. Anaphylaxis
    b. Side effect
    c. Allergic reaction
    d. Antagonism

4. Enteric Coated pills:

    a. May be crushed for ease in swallowing
    b. Dissolves in the stomach
    c. May not be crushed
    d. B and C

5. If a medication has the route of administration listed as "otic," you would administer this medication:

    a. In the eye(s)
    b. In the ear(s)
    c. In the nose
    d. Rectally

6. True or False. Individuals are not permitted to refuse their medications.

Body Substance Precautions
(Universal Precautions)

The concept of Body Substance Precautions presumes that all body fluids are potential carriers of infectious disease and are presumed to be contaminated. Hand washing is the number one technique used in controlling the transmission of infections.

Protective Measures:

(1) Keep your body clean
(2) Practice good hand washing
(3) Don't share personal items such as drinking cups, eating utensils, combs, brushes, etc.
(4) Use trained techniques when working with individuals with communicable diseases
(5) Cover your mouth and nose when sneezing

Hand washing should routinely occur in certain situations:
(This list is not all-inclusive)

- Upon arriving to work or going home at the end of your shift
- After using the restroom
- Before preparing medications
- Before preparing food and after cleaning up from a meal
- After handling any contaminated equipment (urine collection devices, etc)
- Before and after using disposable gloves
- When coming in from an outdoor activity
- After handling any animal
- After sneezing or coughing into your hands

The CDC (Centers for Disease Control) recommends the use of personal protective equipment such as gloves, masks, gowns, eye shields, etc. when there is a possibility of coming into contact with blood or other body fluids. Guidelines for using gloves are:

- Dispose of gloves following approved procedures
- Change gloves before assisting a different individual
- Always wash hands before putting on gloves and after taking the gloves off
- Follow your agency's policy regarding Body Substance Precautions

Waterless hand washing products are available for use in alternative settings when soap and water are not available. They should only be used when there is no access to soap and water. Always follow the instructions with the specific product you are using.

Proper Hand Washing

1. Wet your hands with warm, running water and apply liquid soap (leave the water running).
2. Scrub all surfaces, including the back of your hands, wrists, between your fingers, under your nail beds and around any rings on your fingers. (Actual scrubbing should last at least 20 seconds, which is equivalent to singing the happy birthday song twice).
3. Rinse your hand thoroughly to remove all soap (leave the water running).
4. Dry your hands thoroughly with a disposable paper towel.
5. Turn the faucet off using the paper towel.
6. Use the paper towel when opening the door to exit.
7. Throw the paper towel in the trash can.

Proper Method of Removing Gloves

1. Pinch the palm region of one of the gloves near the wrist. Gently pull the glove off so that it is inside out.
2. Hold the glove in the palm of the remaining gloved hand. Slip two fingers from the ungloved hand under the glove at the wrist of the remaining gloved hand.
3. Pull the glove until it comes off, inside out, so that the first glove ends up inside the gloved just removed.
4. Dispose of the gloves according to your agency's policy.
5. Wash your hands.

MRSA FACT SHEET

MRSA is methicillin-resistant Staphylococcus aureus. This is a potentially dangerous type of "staph" bacteria that is very resistant to certain antibiotics. This type of bacteria can cause skin and other infections. Recognizing the signs and receiving treatment for MRSA when first detected reduces the chances of the infection becoming severe.

MRSA IS SPREAD BY:	Signs and Symptoms of MRSA
Having direct contact with another person's infection	MRSA appears as a bump or infected area on the skin that may be:
Sharing person items, such as towels or razors, that have touched infected skin	* Red * Swollen * Painful * Warm to the touch
Touching surfaces or items which are contaminated with MRSA such as bandages	* Full of puss or other drainage * May be accompanied by a fever

If you suspect MRSA:

Cover the area with a bandage and contact your healthcare professional. Adhere to all Body Substance Precautions taught in this handbook. It is especially important to get in contact with your healthcare provider if there is signs and symptoms of MRSA accompanied by a fever.

Treatment of MRSA:

A healthcare professional may drain the wound and obtain a culture and sensitivity (C & S). The culture that is obtained will be sent to a lab to determine which antibiotics the wound is sensitive to or resistant against. This will allow the doctor to make a decision on which antibiotic to prescribe. NEVER attempt to drain a wound yourself. If antibiotics are prescribed, make sure all doses are taken according to the directions.

Protecting yourself from MRSA skin infections:

- Know the signs of MRSA skin infections and get treatment early
- Keep cuts and scrapes clean and covered
- Encourage good hygiene such as frequent hand washing
- Do not share personal items such as towels or razors
- Always practice Body Substance Precautions

Practice Session

1. The number one thing you can do to help prevent the spread of germs is:

    a. Wear gloves
    b. Cover your mouth when you cough or sneeze
    c. Wash your hands
    d. Wear a face mask

2. In regards to wearing disposable gloves, when should you wash your hands?

    a. Before applying the gloves
    b. After removing the gloves
    c. Both a and b

3. When washing your hands with soap and water, how long should the physical scrubbing portion last?

    a. At least 20 seconds
    b. At least 30 seconds
    c. At least one minute
    d. At least two minutes

4. What should you do if suspect an MRSA wound on an individual supported?

    a. Adhere to the body substance precautions
    b. Cover the wound with a bandage
    c. Set up a doctor's appointment
    d. All of the above

Vital Signs

- Temperature
- Pulse
- Respirations
- Blood Pressure

Important information regarding temperatures

- A normal temperature range is 97.0 to 99.0 Fahrenheit. Report any reading outside this range according to your agencies' policy.

- In order to get an accurate reading you should not take an oral temperature within ten minutes of a person eating, drinking or smoking.

- Unless specifically requested from the doctor, you should avoid using glass thermometers due to the risk associated with them accidentally becoming broken.

- Most digital thermometers no longer require you to add or subtract a degree according to the method used to obtain the temperature. Be familiar with the thermometer you are using.

- If using any device other than a digital thermometer, follow the manufacturer's Instructions and always indicate how the temperature was taken when recording the reading.

Taking an oral temperature using a digital thermometer

- Wash Hands

- Wear disposable gloves

- Cover the thermometer probe with a disposable probe cover

- Press the "on" button

- Wait until the memory clears

- Check the battery (most digital thermometers will flash "lo" if the battery is low

- Insert into the mouth, under the tongue

- Have the client remain as still as possible and instruct them to keep their mouth closed with the thermometer in place

- The thermometer will beep when it has completed recording the temperature

- Remove the thermometer from the individual's mouth

- Note and record the temperature, indicate that the temperature was taken orally

- Remove and discard the probe cover

- Remove the disposable gloves

- Wash your hands

- Report significant deviations to the nurse or doctor

Taking an armpit (Axillary) temperature using a digital thermometer

- Wash Hands

- Wear disposable gloves

- Cover the thermometer probe with a disposable probe cover

- Wipe the armpit region using a dry paper towel

- Press the "on" button

- Wait until the memory clears

- Check the battery (most digital thermometers will flash "lo" is the battery is low

- Place the thermometer in the armpit and have them hold their arm down against their body

- Hold the thermometer in place and instruct the individual to remain as still as possible

- The thermometer will beep when it has completed recording the temperature

- Remove the thermometer from the individual's armpit

- Note and record the temperature, indicate that the temperature was taken in the armpit

- Remove and discard the probe cover

- Remove the disposable gloves

- Wash your hands

- Report significant deviations to the nurse or doctor

Information about Taking a Radial Pulse

- ❖ The radial pulse is located on the underneath side of a person's wrist on the side with the thumb.
- ❖ Place the index and third finger on this location and apply just enough pressure that you can feel the heartbeat.
- ❖ Always count the pulse for one full minute.
- ❖ Note if the pulse is regular, irregular, strong, or weak
- ❖ A normal pulse rate is 60 to 100. Report any pulse less than 60 or greater than 100, according to your agencies' policy.
- ❖ Never use your thumb to take a pulse, you will likely be feeling your own pulse rather than the individual's pulse you are trying to take.
- ❖ A person's pulse is affected by exercise, illness, excitement and medications.
- ❖ A person needs to be sitting relaxed for up to five minutes before taking a pulse.

How to take a Radial Pulse

- Wash your hands

- Using the index and middle finger, find the pulse just below the inside of the wrist on the thumb side

- Count the pulse for one full minute

- Wash your hands

- Record the pulse

- Report any pulse outside of the normal range (60 – 100) to the nurse or doctor

Information about Counting Respirations

- A single respiration consists of a rise and fall of the chest
- Respirations are affected by exercise, excitement, illness and medications
- Count respirations immediately after taking a person's pulse. Keep your fingers on their wrist and count the number of respirations
- Do not tell the person you are counting their respirations or they may alter their respirations unintentionally
- Always count the respirations for one full minute
- Respirations should be slow and easy and between 12 to 20 respirations per minute. Anything outside of this range needs to be reported to a nurse or doctor
- If the individual is in respiratory distress call 911 immediately

Counting Respirations

- Wash your hands

- Take a person's pulse according to the steps listed previously in this handbook

- Immediately after taking a person's pulse, keep your fingers on their wrist and count the number of times the person's chest rises and falls in one minute. Do not tell the person you are counting their respirations

- Wash your hands

- Record the number of Respirations

- Notify a doctor or nurse if the number of respiration in one minute is outside the normal range

Information about Taking a Blood Pressure Reading

- A blood pressure is the force of the blood pushing against the walls of the arteries.
- Taking a person's blood pressure is important. The person <u>may not</u> have any signs and symptoms associated with high blood pressure. High blood pressure can cause damage to blood vessels and increases a person's chance of having a heart attack, stroke or kidney disease.
- Have the individual sit or lay quietly for at least five minutes before taking a blood pressure.
- Do not allow the individual to cross his/her legs while the blood pressure is being taken. Crossed legs increase the blood pressure.
- A woman who has had a breast removed should have her blood pressure taken on the opposite arm.
- You must use the proper size blood pressure cuff or you may get an inaccurate reading. Follow the manufacturer's instructions on verifying that you have the proper size blood pressure cuff.
- Two numbers are measured when measuring blood pressure, a top number and a bottom number. Example: 120/80.
- The higher number, or systolic pressure, refers to the pressure inside the artery when the heart is pumping blood throughout the body. The normal range for the systolic is 90-140.
- The lower number, or diastolic pressure, refers to the pressure inside the artery when the heart is resting and filling up with blood. The normal range for the diastolic is 60-90.
- The use of automated devices are allowed as long as manufacturer's instructions are followed.
- Always carefully read the instructions that come with your blood pressure kit. Each kit may work in a different way.
- Clean the blood pressure cuff and diaphragm of the stethoscope between individuals to help maintain infection control.
- Always clean the ear pieces of the stethoscope before putting them in your ears. Alcohol pads are generally used.
- If you want to double check the blood pressure reading, you need to wait at least three minutes before re-inflating the cuff on the same arm.

Taking a Blood Pressure

- Wash your hands
- Have the Individual sit up or lie down with their arm stretched out. The arm needs to be level with the heart.
- Put the blood pressure cuff (sphygmomanometer) approximately one inch above the elbow. Wrap the cuff snugly around the arm. Make sure the rubber bladder and gauge is centered over the brachial artery. Make sure it is not too tight or too loose.
- Put the ear pieces in your ears.
- Using your middle and index fingers, gently feel for the pulse in the bend of the elbow. This is the brachial artery. You will feel the pulse beating when you find it.
- Place the diaphragm of the stethoscope over the brachial artery pulse. Listen for the heartbeat.
- Tighten the screw the on bulb and quickly squeeze and pump the bulb. This will cause the cuff to tighten.
- Continue squeezing the bulb until the gauge reads approximately 160 or until the gauge reads at least 10 points higher than when you heard the last heartbeat.
- Slowly loosen the screw to release air from the cuff. Let the gauge fall approximately 5 points per second. Carefully look at the gauge and listen to the sounds. Remember the number on the gauge where you first heard the heart beat. This will be your systolic number.
- Continue to listen and watch the gauge, you should continue to hear the heartbeat as the needle on the gauge continues to move down. Note the point on the gauge when you stop hearing the heartbeat. This will be your diastolic number.
- Remove the blood pressure cuff from the person's arm.
- Write down the blood pressure reading, date, time and the arm used.
- Wash your hands.
- Report any reading outside of the normal range to the nurse or doctor.

Practice Session

1. A normal body temperature range for a healthy male would be:

    a. 96.0 to 97.9
    b. 98.0 to 100.0
    c. 97.0 to 99.8
    d. 98.6 to 100.6

2. When taking a pulse:

    a. Use your index and middle finger to apply gentle pressure
    b. Pulse should be felt on the underneath side of the wrist
    c. Pulse should be felt on the side of the consumer's wrist with the small finger
    d. Pulse should be felt on the side of the consumer's wrist with the thumb
    e. A, B and C
    f. A, B and D

3. How long should you take a pulse for?

    a. 15 seconds
    b. 30 seconds
    c. One minute
    d. Two minutes

4. True or False. Before counting a person's respirations, always explain what you are doing before proceeding.

5. Which of the following blood pressure readings should be reported to a nurse?

    a. 92/60
    b. 100/88
    c. 120/80
    d. 144/92

Abbreviations & Measurements

Commonly Used Abbreviations

Abbreviation	Meaning	Latin
AC	Before Meals (at least 30 minutes before meals)	Ante Cibum
PC	After Meals (at least 2 hours after a meal)	Post Cibum
QD	One time per day (Every day)	Quaque Die
BID	Twice Daily	Bis In Die
TID	Three Times Daily	Ter In Die
QID	Four Times Daily	Quarter In Die
QOD	Every Other Day	n/a
PRN	As Needed	Pro Re Nata
Ad Lib	As Desired	Ad Libitum
HS	Bedtime (Hour of sleep)	n/a
HR(S)	Hour(s)	n/a
PO	By Mouth	Per Os
NPO	Nothing By Mouth	Nil Per Os
OD	Right Eye	Oculus Dexter
OS	Left Eye	Oculus Sinister
OU	Both Eyes	Ocular Uterque
Stat	Immediately	Statim
GTT or GTTS	Drop(s)	Guttae

Practice Session:

Transcribe the following physician orders without using any abbreviations.

(1) Amoxicillin 500 mg PO QID X 7 Days

_____

_____

(2) Ciloxan Ophthalmic 0.3% 2 gtts OD x 5 days

_____

_____

(3) Atenolol 50 mg PO QD

_____

_____

(4) Motrin 400 mg PO Every Four Hours PRN Pain or Fever. Not to Exceed Four Doses in 24 HRS

_____

_____

(5) Levothyroxine 100 micrograms PO AC QD

_____

_____

Measurements

1 tsp = 5 ml

1 TBS = 15 ml

1 Ounce = 30 ml

ml = milliliters   tsp = teaspoon

cc = cubic centimeters   TBS = Tablespoon

cc = ml

Practice Session:

(1) How many teaspoons in a tablespoon? _____

(2) How many tablespoons in an ounce? _____

(3) How many teaspoons in an ounce? _____

(4) The doctor orders two ounces of milk of magnesium, how much would you give? _____

(5) The doctor orders ½ ounce of Pink Bismuth, how much would you give? _____

Common Psychiatric Conditions

Depression Fact Sheet

Depression is more than someone simply being unhappy. Clinical depression, also called major depression, is a mood disorder and considered a significant mental health problem. Depression may occur following a life changing event, as the result of a person's general medical condition, or there may be no known cause for the depression. Some types of depression are caused by chemical imbalances in the brain.

Symptoms:

The most common symptom of depression is a sad, depressed mood that:

- Is present most days and lasts throughout the majority of the day
- Lasts for more than two weeks
- Impairs the person's performance at employment, school or in social relationships

Other symptoms of depression may include:

- Loss of appetite
- Weight loss or weight gain
- Sleep problems
- Loss of interest in activities the person generally enjoys
- Withdrawal from friends and family
- Feelings of hopelessness
- Irritability
- Fatigue
- Difficulty concentrating
- Crying easily
- Thoughts of suicide (This symptom needs to be taken seriously)
- Hearing voices (hallucinations) or having strange ideas (delusions)

Treatment:

Treatment varies from person to person. Antidepressant medications are often prescribed. In addition to medications, the person may benefit from counseling, psychoeducation, psychotherapy, and/or electrconvulsive therapy.

Paranoid Personality Disorder Fact Sheet

Paranoid personality disorder (PPD) is a disorder where people often appear odd or peculiar. People with PPD suffer from paranoia. Paranoia is an unrelenting mistrust and suspicion of others, even when there is no reason to be suspicious. The disorder is more common in men and generally begins in early adulthood.

Symptoms:

- Doubt the trustworthiness of others
- Reluctant to confide in others
- Unforgiving and hold grudges
- Read hidden meanings in innocent remarks or facial expressions of others
- Perceive attacks on their character that are not apparent to others
- Recurrent suspicions, without reason, that others are unfaithful
- Cold and distant in relationships
- Believes they are always right and others are wrong
- Difficulty relaxing
- Often hostile, stubborn, and argumentative

People with PPD often believe that others are trying to threaten or harm them. They may stop eating for fear that their food is being poisoned. They often are seen pacing back as forth as they think about statements made during conversations. The more they replay the statements back in their mind, the larger the statement becomes. They often feel that conversations or laughter between innocent parties are about them.

Treatment:

People with PPD often don't seek treatment because they do not feel that they have a problem. Psychotherapy is the choice treatment for PPD. Treatment often focuses on improving general coping skills, social interaction, communication, and self-esteem. Anti-anxiety, antidepressant, or antipsychotic medications may be prescribed if the person's symptoms are considered severe or if the person is also suffering from other associated psychological problems.

Anxiety Disorders Fact Sheet

Anxiety disorders are different from normal feelings of nervousness. Many symptoms may occur including overwhelming panic or fear, uncontrollable obsessive thoughts, intrusive memories, reoccurring nightmares, and physical symptoms such as nausea, increased heart rate, shortness of breath, and muscle tension. People with anxiety disorders often have problems in job performance, school work, and social relationships.

Common Anxiety Symptoms

* Increased heart rate
* Increased Respirations
* Shortness of breath
* Nausea or abdominal pain
* Dizziness or lightheadedness
* Numbness
* Chills or hot flashes

Common types of Anxiety Disorders

- General Anxiety Disorder (GAD) – People with GAD have on-going, severe tension that interferes with daily activities. They often worry constantly and feel helpless to control these worries. Common areas that these people worry about include employment responsibilities, family, health or minor issues such as car repairs, chores or appointments. Anxiety sufferers may have problems with sleep, muscle aches, feeling shaky, weakness and frequent headaches.
- Phobias – This is an excessive and persistent fear of a specific object, situation or activity. These fears cause such distress that some people go to extreme lengths to avoid what they fear. Examples include a fear of heights, spiders, snakes, public speaking, flying, etc.
- Post Traumatic Stress Disorder – This occurs in individuals who have survived a severe or terrifying physical or emotional event. People with PTSD often experience nightmares, intrusive memories and flashbacks. Additional symptoms include irritability, feeling numb, exaggerated startle responses and feeling jittery or on guard.
- Obsessive-Compulsive Disorder – People with OCD are often preoccupied with germs, dirt, nagging doubts and having to have things in a particular order. These people often have rituals that they perform in order to decrease anxiety causing thoughts. An example would include touching an object several times. If the person does not touch the item a certain number of times it increases their anxiety. In order to decrease the anxiety the person must perform the ritual.
- Panic Attack – A panic attack is an overwhelming combination of physical and psychological distress. People experiencing a panic attack often have sudden increases in their heart rate, shortness of breath, sweating, nausea, chills or hot flashes, etc. The symptoms are so severe many people believe they are having a heart attack.

Treatment

Anti-anxiety medications are commonly prescribed to help individuals with these types of disorders. Counseling and Cognitive Behavioral Therapy have also been helpful.

ADHD Fact Sheet

Attention-deficit hyperactivity disorder (ADHD) is usually diagnosed in childhood. Children with ADHD have difficulty paying attention, controlling impulsive behaviors and in general are overly active. From time to time it is normal for children to have difficulty focusing and behaving. In children with ADHD the symptoms continue instead of getting better.

Symptoms of ADHD may include:

- Difficulty paying attention
- Appear to be day dreaming a lot
- Doesn't appear to be listening
- Easily distracted
- Easily forgets things
- Difficulty sitting still
- Fidgets a lot when sitting
- Talks too much
- Not able to play quietly
- Acts and speaks without thinking
- Interrupt others

The first step in determining if a child has ADHD is to set up a medical exam including hearing and vision tests to rule out problems with symptoms similar to ADHD. Another part of the process may include a checklist for rating ADHD symptoms and taking a history of the child from parents, teachers and sometimes the child. Medications may be prescribed to help the child increase his/her ability to concentrate and decrease impulsivity.

Mood Disorder Fact Sheet

Two of the most common mood disorders are depression and bipolar disorder, also known as manic-depressive illness. Researchers believe these illnesses are caused by imbalances in the brain's chemical activity. There is also belief that environmental factors can play a part in mood disorders.

Bipolar Disorder

Description:

Extreme mood swings with periods of generally even-keeled behavior. Bipolar disorder tends to run in families and typically begins in the mid-twenties and continues throughout life. Without treatment, people with bipolar often experience marital breakups, job losses, substance abuse, and suicide.

Symptoms:

When people with bipolar experience mania, they have inflated self-esteem, decreased sleep, rapid speech, increased energy, racing thoughts, feelings of invulnerability, poor judgment, heightened sex drive, and a denial that anything is wrong. When they experience the depressive component they have feelings of hopelessness, guilt, worthlessness, fatigue, loss of appetite, sleep disturbances, and thought of suicide.

Treatment:

There is no lab test that can determine if a person has bipolar. Once the bipolar diagnosis is made medications are used, such as mood stabilizers in conjunction with psychotherapy. Self-help groups can offer emotional support. It is important for people with bipolar to remain on medications. Many people diagnosed with bipolar don't believe they have a problem, and stop taking their medications.

Conduct Disorder Fact Sheet

People with a diagnosis of conduct disorder have difficulty following rules and behaving in a socially acceptable manner. They are often viewed as "bad" or delinquent rather than mentally ill. Many factors contribute to a person developing a conduct disorder, including brain damage, child abuse, genetic vulnerability and traumatic life experiences.

Behaviors commonly displayed by people with conduct disorder

- Aggression to people and animals
- Bullies or tries to intimidate others
- Often initiates fights
- Often uses weapons
- Steals from people
- Forces people into sexual activity
- Property Destruction
- Lies or exaggerates
- Often stays out past curfews
- Runs away

Treatment:

Treatment can be complex and challenging. Behavior therapy and psychotherapy are usually necessary to help the person express and control anger. Treatment may also include medications for those who have difficulty with paying attention, impulse control and/or depression.

Practice Session

Match the psychiatric condition on the left with the selection on the right that is the most appropriate.

____ ADHD   A. Feelings of hopelessness

____ Conduct Disorder   B. Periods of increased pulse and respirations

____ Depression   C. Decreased concentration and increased impulsivity

____ Paranoid Personality Disorder   D. Feeling on top of the world and later very depressed

____ Mood Disorder   E. Thinking people are out to get him/her

____ Anxiety   F. Bullying people, no regards for authority

Diabetes Information

Diabetes

Diabetes is when the pancreas cannot secrete enough insulin to meet the demands of the body or the body cannot effectively use the insulin that is produced. The incidence of Diabetes in the United States has been increasing rapidly. The earlier diabetes is discovered and treated, the fewer complications the individual will experience.

Risk Factors
* Family History (Genetics)
* Inactive lifestyle
* Obesity

Symptoms of Diabetes	
* Hunger	* Frequent urination
* Fatigue	* Unexplained weight loss
* Vision changes	* Wounds that don't heal
* Increased thirst	* Frequent infections

Complications:
- Blindness
- Kidney failure
- Nerve damage
- Amputations
- Infections that don't heal
- Affects every body organ

Prevention of Complications
- Routine exercise
- Well balanced diet
- Maintaining a healthy weight
- Taking medications as ordered
- Monitor blood sugars as ordered
- Inspect skin, especially the feet routinely
- Wear shoes or slippers to prevent foot injuries

Recommendations for People With Diabetes
- Yearly eye exams and six month dental check-ups
- Blood pressure screenings routinely, EKG as ordered by the physician
- Eat three healthy meals per day with healthy snacks between meals
- Only have a health care professional cut toenails
- Obtain routine blood work as ordered by the physician

Blood sugar readings should always be taken before a person eats or drinks anything in order to get a valid reading. A fasting blood sugar reading should fall between 70 and 110. Any reading below 70 is considered hypoglycemia and any reading above 120 is considered hyperglycemia.

When you eat food, the food enters your stomach and the stomach begins to break the food down. The sugar that is released into the blood stream cannot be used by the body. Your pancreas has to secrete insulin in order to break the sugar down into the only usable form of sugar that the body can use (glucose).

If our bodies do not secrete enough insulin, this un-usable form of sugar circulates throughout our bodies and blood sugar levels rise. As the blood sugar rises, our body attempts to get rid of this sugar by causing thirst. The goal is to dilute out the sugar and then excrete the sugar in the urine. With increased thirst comes the increased urination, not only due to the increased fluid being taken into the body, but also because the sugar is irritating to the bladder and stimulates the feeling of having to urinate often.

Signs and symptoms of hyperglycemia include:

- Feeling weak
- Tired or drowsy
- Stomach upset
- Nausea / vomiting
- Confused or disoriented
- Fruity smelling breath
- Rapid respirations

Treatment includes encouraging zero calorie fluids and regular exercise (getting up and moving around) to help decrease the blood sugar level. If the person has medication orders, you will need to follow the orders as outlined by the physician. Only caregivers that have completed an approved Insulin Administration Course may administer insulin. Follow the doctor's guidelines as to when the person needs to be taken to the emergency room. In general, a blood sugar reading of 300 or above (unless indicated otherwise by the physician) needs medical attention. Call 911 if the person becomes unconscious.

Hypoglycemia is when a person's blood sugar level drops below 70. This often occurs when people experience stress, illness, increased activity level, or when they do not take in enough food. Hypoglycemia is generally easily reversible.

Signs of Hypoglycemia

- Nausea
- Headache
- Increased pulse rate
- Hunger
- Blurry vision
- Unsteadiness
- Tingling in hands and face
- Tremors / shakiness
- Dizziness / lightheadedness
- Irritability

Treatment for Hypoglycemia:

(1) Obtain a blood sugar reading less than 70.
(2) If the person is alert and oriented, give the person approximately 15 grams of a rapid acting sugar (carbohydrate) which should raise the blood sugar by 30-45 points (see examples on the next page).
(3) Stay with person and recheck the blood sugar 15 minutes after administering 15 grams of the fast acting carbohydrate.
(4) If the blood sugar after 15 minutes remains lower than 70, repeat step three. If the blood sugar level is above 70 the person will need to eat carbohydrates and protein in order to maintain the blood sugar level within the normal limits. Examples of these types of food include: crackers and peanut butter, crackers and cheese, one cup of milk.
(5) If at any point the person becomes unconscious, call 911. Never put anything in a person's mouth if they are unconscious or disoriented/confused. This increases the risk of the person choking or aspirating.

Hypoglycemic Carbohydrate Examples
Each item is equal to 15 grams of a fast acting carbohydrate
6 Life Savers chewed and swallowed
8 ounces of milk
6 ounces of juice
Medium sized fruit
1/2 can of non-diet soda
4 glucose tabs
1 tube of glucose gel
1 tablespoon honey, jelly or jam
1 tablespoon pancake syrup
1/2 of a regular candy bar

If at any point the individual becomes:

- **Increasingly symptomatic**
- **Less alert or loss of consciousness**
- **Experiences difficulty breathing**
- **Has a seizure**

Call 911 Immediately

Glucagon Kits:

If a glucagon kit has been prescribed, <u>the glucagon kit may be administered as long as the person administering it has received specialized training</u>. Glucagon should only be administered for a person with a blood sugar less than 70 when the person is either unconscious or it is not safe to put food or liquids in the person's mouth due to the risk of choking. The order for the steps would be (1) Administer the Glucagon (2) Call 911. When glucagon is injected it releases glucose that is stored in the liver in order to raise blood sugar levels quickly.

Managing Diabetes Through Proper Nutrition

Foods to Avoid

White breads	Cakes / Pies
White rice	Candies
White potatoes	Cookies
White pasta	Donuts
Regular Soda	Juices
Noodles	Chips

Good Food Choices

Whole wheat breads, rice, pastas	Sugar free Jello
Low fat popcorn selections	Fruits
Reduced calorie breads	Vegetables
Low fat food selections	Lean Meats
Artificially sweetened desserts	High fiber foods
Diet versions of sodas	Low fat Milk

Good Bedtime Snacks

String Cheese
5 saltine crackers with peanut butter
6 ounce glass of milk
1 oz of nuts (almonds or peanuts)

Daily Requirements

No more than 180-200 carbohydrates per day
Approximately 25 fat grams per day
Approximately 50-60 grams of protein per day
At least 30 grams of fiber per day

There needs to be a good balance between carbohydrates, protein, and fat. Carbohydrates have the greatest effect on a person's blood glucose level, where protein and fat have less of an effect. A person with diabetes should eat three small meals a diet with appropriate snacks between meals. Snacks help to decrease fluctuations in blood sugars by not going too long without food. People with diabetes can eat anything that anyone else can eat, but they need to eat high sugar food in moderation. Always watch portion sizes and read nutrition labels carefully.

Healthy Eating tips

- Do not skip meals
- Eat regularly and space carbohydrates evenly throughout the day
- Meals should include a good source of carbohydrate, lean protein and/or healthy fat
- Include 1 ½ cups vegetables, 2-3 servings of fruit, 1-3 cups of low fat milk or yogurt
- Drink water or zero calorie drinks. Goal is 64 ounces per day (Eight 8 oz glasses)

Glucometer Training

- A glucometer is a device used to check a person's blood sugar.
- Blood sugar readings should be taken before a person eats or drinks anything.
- Supplies needed:
    - Glucometer Machine
    - Gloves
    - Lancet
    - Sharps Container
    - Alcohol pads
    - Test Strip for the glucometer machine
    - Lancet Pen or other device

Steps:

1. Assemble equipment
2. Identify the individual and explain the procedure
3. Wash hands and apply gloves
4. Place lancet in a lancet pen (if using a device with the lancet)
5. Set up the glucometer
6. Clean individual's finger with an alcohol pad
7. Turn glucometer on, apply lancet to the side of the finger (never the finger pad)
8. Point finger downward and gently squeeze to get an adequate blood sample
9. Place drop on the end of the testing strip and wipe finger with alcohol pad or guaze
10. Read and record the result on the medication administration record
11. Clean equipment and dispose of lancet in sharps container
12. Remove gloves and wash hands

Note:

Some devices allow you to take blood sugar readings from sites other than the fingers. Follow the equipment's recommendations on where blood sugar readings can occur. Always use the fingers unless instructed otherwise.

Rotate fingers to avoid the formation of calluses.

Once a month the machine must be checked using either a "normal" solution or "High and low" solutions. These checks are to verify the machine is working correctly. Follow the recommendations with the solutions and the machine you are using. Always document the "normal" reading or "highs and lows" on the MAR. Do not use the machine if the controls are not within their correct ranges.

Some machines require you to calibrate the machine every time a new box or bottle of test strips is opened. Always read the information that comes with your machine and follow all directions in order for the machine to give accurate results.

Practice Session

(1) What is Diabetes?

_____
_____
_____

(2) Outside of Medication how is Diabetes Managed?

_____
_____

(3) In regards to Diet, what foods should be encouraged and what foods should be used sparingly?

_____
_____
_____

(4) What are some examples of healthy snacks a diabetic can eat?

_____
_____

(5) If a person's blood sugar is low (less than 70) and they have tremors and obvious signs of Hypoglycemia, but they are alert and oriented, how would you help them get their blood sugar level up?

_____
_____
_____
_____

(6) Name two signs and symptoms of hypoglycemia (Fasting blood sugar below 70).

_____
_____

(7) Name two Signs and symptoms of Hyperglycemia (Fasting blood sugar above 110).

_____
_____

(8) What would you do for a person that had a blood sugar reading of 30 and you cannot wake him up?

_____
_____
_____

Seizures

Seizures

Seizures occur when there is abnormal electrical activity (or "misfiring") in the brain. This causes a wide range of behaviors depending on where the "misfiring" occurs. Seizures cause some people to fall to the ground with uncontrolled "jerking" movements. Other seizures cause the person to stare off into space for a period of time.

Most people do not die directly from a seizure. They can die from becoming injured during a seizure such as hitting their head or drowning while in a bathtub or swimming pool. People who "stare off" during a seizure could cause an automobile accident if driving.

Many people have warning signs before having a seizure. This is called an aura. Auras are specific to the person but may include things like a smell, headache, hearing certain sounds or seeing a flashing light, etc.

Most seizures are short and last less than five minutes. People are generally tired and sometimes confused after a seizure ends (this is referred to as the "postictal period"). They should be allowed to rest. Individuals often recover within 20 to 30 minutes after a seizure, but this varies from person to person.

Two things must be done when you witness a person having a seizure.

(1) Keep the person safe. Move objects and furniture out of the way of the individual. If an object cannot be moved, try placing a pillow or anything soft between them and the hard object. Try not to move the person if at all possible. If they are in bed try to prevent them from falling out of bed, assist them to the ground if they are close to the edge and it looks like they are going to fall out of bed. Do not put anything in their mouth. If possible, position the person on his/her side to prevent aspiration and maintain the airway.
(2) Immediately start timing the seizure. Use the second hand on a watch or wall clock in order to get an accurate time on how long the seizure lasts.

You need to be aware of the person's seizure protocol if they have a history of seizures. This will be located in the person's Individual Plan and will give information about what a typical seizure looks like for this individual along with step by step instructions on what needs to happen during the seizure. Don't wait until a person has a seizure to read the protocol, be familiar with it so you are prepared.

If a person has never had a documented seizure you will need to call 911 immediately when the seizure starts.

If a person with a seizure history has a seizure that lasts longer than five minutes, 911 should be called <u>unless</u> you have specific written orders signed by a Neurologist detailing when 911 should be notified.

Sample Seizure Protocol:

SEIZURE PROTOCOL FOR JOHN DOE

- Move all items away to reduce the risk for injury. Do not restrain the consumer or put anything in his mouth.

- Start timing the seizure immediately.

- Monitor for respiratory distress (difficulty breathing, blue-gray color of nail beds or lips), or aspiration/choking on a foreign object (food, vomit, etc).

- **Call 911 if JOHN has a tonic-clonic seizure (muscle contractions and loss of consciousness) that lasts more than _____ minutes.**

- **Administer 1mg of Ativan (Lorazepam) sublingual if the seizure lasts 3 minutes or longer.**

- Call the neurologist any time the CONSUMER is taken to the ER or any time that the CONSUMER has a change in his pattern of seizures.

- Notify the supervisor and nurse following all seizures.

- Complete the seizure log.

- Complete an incident report if an injury occurred as a result of the seizure.

_____   _____

Physician Signature   Date

Precautions for People With Seizures
^ When taking a shower or bath always have someone in close proximity to help in the event of a seizure
^ Avoid getting the water too hot or cold while bathing as this could set off a seizure
^ Wear a life jacket when swimming and **NEVER** swim alone

Causes of Seizures
Head Injuries
Lack of oxygen during birth
Brain tumors
Genetic conditions
Lead poisoning
Prenatal brain development problems
Infections: meningitis or encephalitis
7 out of 10 have no known cause

Potential Seizure Triggers
* Flickering lights
* Bright sun light
* Not getting enough rest
* Change in temperature (too hot or cold)
* Increased stress
* Loud noises
* Not taking medications as prescribed
* Sometimes no known triggers

When To seek Emergency Medical Assistance

The person stops breathing
Seizure lasts longer than specified by the physician
The person sustains an injury during the seizure
One seizure follows another without a break
Person remains unconscious after the seizure
This is the first seizure a person has had
Seizure occurred in water
The person having a seizure is pregnant

Common Seizure Types
Generalized Tonic-Clonic Seizure (Grand-Mal): These seizures begin with stiffening of the limbs (tonic phase) followed by jerking of the limbs and face (clonic phase)
Myoclonic Seizures: These are rapid, brief contractions of body muscles. They generally occur at the same time on both sides of the body but they can involve just one limb, hand or foot
Atonic Seizures: These produce an abrupt loss of muscle tone. They are also referred to as "drop" seizures, astatic ,or akinetic seizures. They can produce head "drops", loss of posture, or sudden collapse.
Absence Seizures (Petit Mal): These are lapses of awareness, sometimes with staring. They generally last only a few seconds and occur without warning.
Partial Seizures: These are divided into simple partial (consciousness is maintained) or complex partial (consciousness is impaired). Virtually any movement, sensory, or emotional symptom can occur as part of a partial seizure, including complex visual or auditory hallucinations.

Seizure Logs

Seizure records need to be maintained on all individuals with seizure histories. Seizure records must be completed for every known seizure the individual experiences. These logs help both staff and the neurologist look for patterns or changes in the individual's seizures. There may be a change in the type of seizure the person is having or the length of seizure. Patterns can also help identify activities or situations that may be able to be changed without requiring medication changes. For example, seizures that occur during showers may indicate the water is too hot, or seizures that occur during specific activities may indicate that the activity is too stressful.

The following is an example of a seizure record. It is important that every seizure be recorded and that several months worth of these records are brought to every neurologist appointment so they can be reviewed.

Name _John Doe_   Month _September_   Year _2011_

A. Tongue Biting
B. General body stiffening
C. Jerking of entire body
D. Staring
E. Head dropping
F. Unconsciousness
G. Right side jerking
H. Left side jerking
I. Loss of muscle tone
J. Incontinent of urine
K. Incontinent of stool
L. Eyes rolling back
M. Picking at clothes
N. Other – describe

Date	Time	Duration	Description	Location
9/8/2011	7:35 PM	35 seconds	A. M. J.	Living Room
9/12/2011	8:45 PM	42 seconds	G. A. J. I.	Bathroom

Treatment of Seizures with DIASTAT

Diastat is a medication that is administered rectally while the person is still having a seizure. Some people have prolonged seizures that may be hard to stop and emergency treatment may be indicated to stop the seizure and prevent brain damage.

Diastat is a gel formulation of Diazepam. It is administered rectally for treatment of serial or cluster (repetitive) seizures or seizures that last longer than specified by the physician. It can only be ordered by a physician.

Seizures generally stop within 5 to 15 minutes after receiving the medication. If the seizure continues, follow the seizure protocol or call 911 for help.

The medication needs to be available to the individual in the event he/she has a seizure. It will be necessary to carry this medication with you whenever transporting the individual into the community. The medication cannot get too hot or too cold. The medication is viable for three years so you need to be aware of the expiration date.

Call 911 if . . .

- Seizure lasts longer than 15 minutes after giving Diastat
- Seizure behavior is different from other episodes
- You are alarmed by the frequency or severity of the seizure
- You are alarmed by the color or breathing of the individual
- The individual is having unusual or serious problems

Steps for Administering Diastat

(1) Place the individual on their side in a location where they cannot fall

(2) Get the medication

(3) Put on gloves

(4) Get the syringe from the package

(5) Remove the protective cover from the syringe with a downward firm pulling away motion

(6) Lubricate the rectal tip with lubricating jelly in the package

(7) Turn the individual toward you and pull down clothing to expose the buttocks

(8) Bend their upper leg forward to expose the rectum

(9) Separate the buttocks to expose the rectum

(10) Gently insert the syringe tip into the rectum

(11) Slowly count to three while pushing the plunger until it stops

(12) Slowly count to three again before removing the syringe from the rectum

(13) Slowly count to three while holding the buttocks together to prevent leakage

(14) Keep the individual on their side facing you. Note the time the Diastat was given and continue to observe.

Vagus Nerve Stimulators

A vagus nerve stimulator is a medical device approximately the size of a half dollar. It is surgically implanted in a patient just under the skin, on the left side of the body above the heart. You can easily feel the device by running your hand over the upper chest of the person. This device has wires that are surgically attached to the vagus nerves. Throughout the day it is programmed to give stimulation to the vagus nerves toprevent seizures from occurring.

VNS's are used when a patient has experienced multiple drug failures in trying to prevent seizures. A person with a VNS will also typically be on seizure medications at the same time. The doctor is the only one that changes the settings on the device, and changes are done in the doctor's office. The doctor will use an external device in his office during routine visits to check the settings, verify the battery strength, and change the settings if desired.

If a person with a VNS experiences an aura, he/she will swipe a designated magnet directly across the VNS devise. In many cases this will prevent the seizure from occurring , or will lessen the strength of the seizure.

If you are caring for a person who has a VNS device, you should be familiar with where the special magnets are located in the home. Typically the magnets are kept on the front of the refrigerator for the home. A magnet also needs to be carried with the person when they go out into the community.

If a person experiences a seizure you need to get the magnet and swipe it across the VNS directly against their skin. Only swipe it the number of times indicated on the seizure protocol. Continuously swiping the magnet will not increase the stimulation because the device is only programmed to send extra stimulation one time for the event. Continuously swiping the magnet will cause the battery to weaken and the only way to replace the battery is with surgery. A typical battery lasts 3-5 years depending on the settings.

Side effects of VNS therapy are generally mild and temporary. Potential side effects can include frequent coughing, clearing of the throat, or itching on the left side of the individual's neck. These side effects are often reduced by adjusting the level of stimulation.

Please refer to information provided by the neurologist for further information about the VNS device the individual is using.

Benefits of Vagus Nerve Stimulators
Helps prevent seizures or reduce seizure severity
Enables a person to reduce the number of anti-seizure medications needed
Provides long-term quality of life benefits in addition to seizure control

Practice Session

1. A person without a known seizure disorder experiences a seizure. When should you call for emergency medical personnel?

    a. Only if the seizure lasts greater than 3 minutes
    b. Only if the seizure lasts greater than 5 minutes
    c. Immediately
    d. Monitor them for respiratory distress, if the seizure stops under three minutes, set up a routine appointment with the person's doctor to discuss the seizure

2. Which of the following may trigger a seizure in a person with a known seizure disorder?

    a. Flickering lights
    b. Not getting enough rest
    c. Loud noises
    d. All of the above

3. True or False. Seizure logs are only required for people who experience six or fewer seizures per month.

4. Which of the following is true in regards to the medication Diastat?

    a. Diastat is administered orally during a seizure according to the seizure protocol
    b. Diastat can only be administered by an LPN or RN
    c. Diastat should not be administered while out in the community, it should only be administered in the privacy of one's home
    d. None of the above

Drug Reference Materials

Drug Reference Material

Before administering any medication, you need to know about the medication. You should be familiar with what the medication is used for, proper dose ranges, and side effects that you should be observing for.

Independent Supported Living and Group Homes need to have side effect sheets placed with the Medication Administration Records. These sheets serve as a quick reference for the person passing medications (instead of searching through a drug reference book for every medication). These sheets can be printed from various trustworthy websites or they may be obtained from the pharmacy when prescriptions are picked up.

It is also suggested that a drug reference book be available in the home. Having internet access available helps people working with individuals so that medications can be looked up if there are questions or if side effect sheets are missing. The best practice is to look medications up from the manufacturer's web site.

Drug reference manuals are a good resource for the person passing medications, for the individual supported to learn about his/her medications, for nurses working with the individual, and for anyone else working with the individual who would like to learn about the medications prescribed.

Information contained in a drug reference book includes:

- Brand names and generic names of a particular medication
- Available forms the medication comes in (liquid, tablets, capsules)
- Indications and doses (what the medication is typically prescribed for along with dosage information according to the indication)
- Administration (guidelines for administering the medication by all applicable routes)
- Action (how the medication provides it therapeutic affect)
- Adverse reactions (common adverse reactions experienced by 10% of people taking the medication in clinical trials)
- Interactions (a list of each medication's confirmed, clinically significant interaction with other medications)
- Effects on lab test results (lists increased or decreased levels, counts, and other laboratory test results, which may be caused by the medication's systemic effects)
- Contraindications (lists any conditions in which the use of this drug is undesirable)
- Nursing Considerations (useful information on monitoring techniques, and suggestions for prevention and treatment of adverse reactions)
- Patient teaching (guidelines for explaining the drug's purpose, promoting compliance, ensuring proper use and storage, and preventing or minimizing adverse reactions).

Practice Session:

Using a drug reference book answer the following questions:

(1) How long should you take a pulse before administering the medication Digoxin?
    a. 30 seconds
    b. 60 seconds
    c. 2 minutes
    d. Not indicated for this medication

(2) Which of the following medications is contraindicated if you are taking Warfarin?
    a. Zoloft
    b. Tylenol
    c. Aspirin
    d. Atenolol

(3) When is the best time to administer Lasix?
    a. First thing in the morning
    b. 30 minutes before supper
    c. Take with food during supper
    d. At bedtime

(4) Which of the following medications are used to help control seizures?
    a. Sertraline
    b. Albuterol
    c. Depakote
    d. Docusate Sodium

(5) What is a potential adverse reaction from the medication Zyprexa?
    a. Diarrhea
    b. Weight loss
    c. Hair loss
    d. Tremors

(6) Why would someone take the medication Celexa?
    a. Seizures
    b. Diabetes
    c. Depression
    d. Hypertension

Medication Classifications

Medication Classifications

Category	Category Defined	Examples
Alzheimer	Slows down the progression of the disease	Aricept, Namenda
Analgesics	Used to control pain	Aspirin, Tylenol, Percocet
Antacids	Used to control heart burn/indigestion	Pepcid, Tagamet, Tums
Anti-Anxiety	Calms the nerves, helps control anxiety	Xanax, Valium, Lexapro
Antiarrhthimics	Used to regulate an irregular heart beat	Toprol, Sotolol
Antiasthmatics	Used to treat asthma	Albuterol, Proventil
Antibiotics	Used to treat bacterial infections	Penicillin, Amoxicillin, Cipro
Anticonvulsants	Used to treat seizure disorders	Dilantin, Tegretol, Depakote
Antidepressants	Used to treat depression	Zoloft, Paxil, Lexapro
Anti-diarrheals	Used to stop diarrhea (watery stools)	Imodium, loperamide
Antihistamines	Used to treat allergic reactions	Claritin, Benadryl, Zyrtec
Antihypertensives	Used to control blood pressure	Atenolol, Verapamil
Antiparkinsons	Used to control tremors	Sinemet, Levadopa
Antipsychotics	Used to control hallucinations, delusions	Zyrexa, Abilify, Seroquel
Contraceptives	Used to prevent pregnancy	Yasmin, Depo Provera
Diabetic Drugs	Used to treat diabetes	Avandia, Metformin
Diuretics	Used to get rid of excess fluid (water pill)	Lasix, Aldactone
Expectorants	Thin secretions and promote coughing	Mucinex, Humibid
Hormones	Used to treat hormone related conditions	Levothyroxine, Estrogen
Laxatives	Used to stimulate bowel movements	Miralax, Colace, Senna
Mood Stabilizers	Used to even out erratic mood swings	Lithium, Depakote
Nutritional Supplements	Used to make up for deficiencies in vitamins and minerals	Multivitamins, Ensure, Boost, Centrum, Caltrate
Osteoporosis	Used to increase bone density	Fosamax, Boniva, Actonel
Psychotropics	Any drug used to treat mental illness	Any drug used for this purpose
Sedatives/Hypnotics	Used to promote rest and sleep	Lunesta, Ambien, Benadryl
Skeletal Muscle Relaxants	Used to control muscle spasms and pain from injury	Baclofen, Flexeril, Skelaxin, Norflex
Stimulants	Calms individuals with ADHD or ADD	Ritalin, Adderall, Caffeine

General Information About Each Medication Classification

Analgesics
These are pain medications. They often cause stomach upset, so give with food/milk unless directed otherwise.

Antacids
May cause constipation. Can interfere with the absorption of other medications. Best to give one hour before or two hours after other medications.

Anti-Anxiety
Do not stop these mediations suddenly. The person will need to be weaned off slowly. Avoid alcohol consumption. Signs of overdose include: unsteady gait, Drowsiness, and slurred speech.

Antiasthmatics
If the medication causes stomach upset, give with food (unless contraindicated). Heart rate may increase, monitor the pulse after administering the medication. Report heart rates above 100 to a nurse.

Antiarrhthmics
Generally you check the pulse before administering these medications. Medication Administration Record will tell you when to hold medication if the pulse is too low (Usually < 60). Notify the nurse or doctor if individual's heart beat is irregular or too slow.

Antibiotics
Potential for allergic reaction or other adverse reactions such as diarrhea or nausea. If the order reads every 8 hours it must be every 8 hours (7am, 3pm, 11pm). This will keep blood levels steady to help fight off infections. Take all of the medications prescribed, stopping early can lead to antibiotic resistant infections.

Anticonvulsants
Read instructions closely regarding whether the medication should be given with or without food.

Antidepressants
Watch for signs of depressions, which includes signs for suicide. Be aware of any dietary restrictions associated with the specific medications the individual is taking. Photosensitivity may be a side effect, if so wear sunscreen and monitor the amount of time in the sun (including in the winter time). These medications can take several weeks before the full therapeutic effects are seen. Dry mouth can be associated with these types of medications. Keep hard candies and ice chips available or other alternatives according to the individual's specific dietary needs/restrictions. If a person is drinking over 64 ounces of fluid in a 24 hour period notify the nurse or doctor. Water intoxication can occur if a person drinks too much fluid.

Anti-Diarrheals
Report diarrhea to a nurse, according to your agency's policy.
Clean the rectal area well after each stool to avoid skin breakdown.

Record color, odor, consistency and number of stools when charting.
Give a bland diet as long as diarrhea persists

Antihypertensives
Blood pressure should be monitored regularly with parameters in place for when to call a nurse.
Check for dietary restrictions with these

medications.

Antihistamines
May cause drowsiness and dry mouth.
May cause decreased concentration and motor
control in some people.

Antiparkinsons
Increased risk of falls while taking these medications.

Antipsychotics
Many of these medications have weight gain as side effect along with the possibility of developing diabetes.
Be aware of Neuroleptic Malignant Syndrome:
Fever, rigid muscles, sweating, renal failure, muscle wasting, pallor.
Photosensitivity is seen with many of these medications. Wear sunscreen even in the winter and avoid long periods in the sun.

Contraceptives
Report any leg pain or visual problems to the nurse or doctor.
Smoking increases the risk of blood clots.
Some medications decrease the effectiveness of these medications if taken together. This includes
some antibiotics.
Do not miss any doses of the prescribed medication. Missing doses can result in an increased risk for pregnancy.
Prescribed contraceptives do not protect the person against sexually transmitted diseases.

Diabetic Medications
Always monitor a person's blood sugar as ordered.
Infections and antibiotic use can increase a person's blood sugar.

Diuretics
Monitor weight and Input/output as ordered.
Observe for swelling in ankles, legs and hands.

Expectorants
Drink plenty of fluids with these medications.
Deep breathing and coughing as instructed.

Laxatives
Extensive use can lead to laxative dependence.
Drink 6-8 eight ounce glasses of water daily.
Daily fiber intake should be a min. of 25 gms.

Mood Stabilizers
Report signs of toxicity: Lethargy, imbalance, slurred speech, nausea, ringing in the ears, etc.
Many require lab tests routinely.

Practice Session

Match the mediation classification on the left with the most appropriate definition on the right.

____ Analgesics   A. Used to control hallucinations and delusions

____ Diuretics   B. Calms people with ADHD

____ Psychotropics   C. Used to treat seizure disorders

____ Stimulants   D. Used to controls tremors

____ Anti-Psychotics   E. Used to control pain

____ Antiparkinsons   F. Used to treat mental disorders

____ Anticonvulsants   G. Used to get rid of excess water

Adverse Reactions

Anaphylaxis

Epi-Pens

Adverse Reactions

Many factors are taken into account when a medication is prescribed for an individual. The physician or advanced practice nurse must be made aware of all medications the person is receiving, including over-the-counter medications (OTC).

Factors Affecting Medication Dosage and Action	
Age	60 year old will likely receive less of a dose compared to an 18 year old
Weight	Most medication doses are based on a person's weight, for example, Depakote's maximum dose is 60mg/kg
Sex	Certain medication are generally prescribed only for males or only for females
Allergies	The person should not receive a medication that they are allergic to
Activity Level	The more athletic a person is, the more efficient his/her body is at using the medication
Other Medications	The physician needs to be aware of all medications the person is receiving in order to avoid medication interactions

Other Considerations	
Form of the Medication	Does the medication need to be liquid for someone who has a hard time swallowing? Extended release to avoid giving a medication multiple times per day?
Time	Does the medication make the person drowsy? Is the person experiencing a behavior at a certain time when medications could offset this behavior?
Route	If the person is experiencing nausea and vomiting, it may be necessary to receive a suppository instead of an oral medication.
Interactions with food	Some medications cannot be administered with certain foods. Grapefruit is one example.
Interactions with other medications	Some medications cannot be given at the same time as other medications. Antacids are contraindicated with several medications but may be given two hours later.

A drug tolerance occurs when an individual's body has adapted to the current dose of a medication. The dose has to be increased in order for the individual to receive the desired effect that the previous dose once produced.

Cumulative effect is a buildup of the medication in the person's body. Some medications such as fat soluble vitamins are stored in the fat instead of being excreted in the urine like water soluble vitamins. Over time the person may receive too much of the medication and display signs of toxicity or poisoning.

Drug interactions occur when drugs react to each other within the body. Some of these interactions are antagonistic, which means they work against each other. In this case you would get less of an effect of one of the two medications or less of an effect of both medications. Another reaction that may occur is raising the effects of another medication, often beyond what is desired. Drug interactions could also increase laboratory values of some medications.

When a person takes a medication, whether it is a new medication or a medication they have taken for a long time, they need to be observed for possible adverse reactions. Techniques of observation include:

- Sight – Physical things you see that may be occurring with the individual: changes in the way the person is walking, changes in the color of their stool or urine, emesis, rashes, hives, bruises, etc.
- Hearing – Things that you hear such as wheezing, coughing, complaints about the way they feel, moaning, or even silence.
- Smell – When people have elevated blood sugars, their breath will smell "fruity" or sweet. There may also be abnormal odors related to the urine, stool or emesis.
- Touch – The person's skin may feel warm, cold or wet or they may have an exaggerated response to being touched.
- Vital sign changes – Increase or decrease in the pulse, blood pressure and/or respirations. Temperature may be affected, but generally this is not an initial response.

Any medication, (prescription or over the counter), may cause an adverse reaction

A person can become allergic to a medication even if they have taken the medication in the past without any problems. You can even wake up one day and be allergic to a medication that you have taken daily for years. Over time our body chemistry changes and this can lead to new medication and food allergies.

Common Symptoms of Adverse Reactions:

- Difficulty breathing
- Wheezing
- Choking
- Skin turning blue
- Rash
- Hives
- Itching
- Swelling of the throat
- Nausea
- Vomiting
- Sweating
- Changes in vital signs
- Dizziness
- Behavior changes

The above adverse reactions are considered medical emergencies and the individual needs to receive emergency medical care.

Anaphylaxis

- Anaphylaxis is a sudden, severe, potentially fatal allergic reaction. Symptoms generally occur within minutes to hours after receiving an allergy-causing substance.

- Common causes of anaphylaxis include:
    - Foods such as shellfish or nuts
    - Medications such as penicillin or sulfa
    - Bee stings
    - Latex

- Common symptoms of anaphylaxis:
    - Hives
    - Difficulty breathing or swallowing
    - Swelling of the tongue, throat, lips, or around the eyes
    - Loss of consciousness
    - Increased heart rate
    - Metallic taste or itching in the mouth
    - Increased anxiety
    - Rapid decline in blood pressure
    - Generalized itching or redness of the skin
    - Abdominal cramps and/or diarrhea
    - Nausea and/or vomiting

- How to respond appropriately:
    - If an Epi Pen has been ordered, use it according to your training
    - Call 911
    - Perform rescue breaths and/or full CPR according to your training/certification (such as American Red Cross or American Heart Association)
    - If an Epi Pen was administered, do not discard the Epi Pen. This will be taken with the individual to the Emergency Room. The individual may seem to recover after the injections, but they still need to be assessed by medical personnel. There may be a reoccurrence.

- The three "R's" associated with treating anaphylaxis are:
    - Recognize the symptoms
    - React quickly
    - Review the episode and attempt to prevent it from reoccurring

General Information and instructions on using an Epi-Pen:

- An Epi-Pen is used to treat severe allergic reactions.
- People who are allergic to bee stings may need to carry an Epi-pen with them at all times.
- In an emergency the auto injector pen may be used through clothing.
- Do not expose the pen to direct sunlight or excessive heat or cold.
- Make sure the pen is not expired, have a system in place for periodically checking the date and replacing it when needed
- <u>Epi-pens are a nursing delegated task. A nurse will need to provide additional training to you before you can administer an Epi-Pen.</u>

Steps for using an Epi-Pen:

1. Put on gloves
2. Pull off the safety cap on the Epi-Pen
3. Place black tip on outer thigh (it can only be applied to the outer thigh)
4. Using a swinging motion, jab the needle into the thigh
5. The medication from the syringe should automatically go into the person. Count to ten before removing the needle to make sure all of the medication went into the thigh
6. Massage the area for ten seconds to help speed up the absorption of the medication
7. Call 911, the effects of the medication only last approx. 15 minutes
8. Save the Pen used and give this to the ambulance personnel when they arrive
9. The person needs to be taken to a hospital

Practice Session

1. True or False. If a person has been taking a medication for a long time, you don't have to be concerned about a possible allergic reaction with this medication.

2. True or False. Over-the-counter medications don't require monitoring for adverse reactions compared to prescription medications.

3. What are the five techniques of observation following a medication being administered?

    a. _____

    b. _____

    c. _____

    d. _____

    e. _____

4. Which of the following is not a symptom of an adverse reaction?

    a. Skin turning blue
    b. Difficulty breathing
    c. Rash
    d. All of the above are signs of an adverse reaction

5. True or False. An Epi Pen is the treatment for a severe allergic reaction. A properly trained staff may administer the injection and there is no need to call for Emergency Medical Personnel.

Psychotropic Medications

Psychotropic Medications

Psychotropic medications are prescribed to help improve a person's mental health, their behavior, or symptoms of mental illness. Examples include anti-depressants, antipsychotics, mood stabilizers, anti-anxiety agents, sleep agents, stimulants, and anti-parkinson medications. If used for psychotropic reasons, anti-convulsants and cardiac medications would also be considered psychotropic.

Observation and Reporting of Side Effects

People taking psychotropic medications may be unable to verbalize symptoms or medication side effects. It is important for the caregiver to watch for signs of possible side effects and to report them to the nurse or doctor.

Minor Side Effects Common to Psychotropics:
- Dry Mouth
- Change in sex drive
- Increase in thirst
- Drowsiness
- Photosensitivity

Serious Side Effects to Report Immediately:
* Allergic Reactions (difficulty breathing, swelling, rash, fever, etc.)
* Excessive sleepiness, insomnia, confusion or disorientation
* Change in stools (diarrhea or constipation)
* Irregular heartbeats, heart rate outside of normal limits
* Change in blood pressure (too high or too low)
* Fainting or dizziness, especially with position changes (sitting to standing)
* Abnormal posture, movements or gait
* Yellowing of the eyes or skin
* Unusual bruising or bleeding

There may be an increased risk for people taking antipsychotic medications of developing diabetes. As a precaution, many doctors will obtain baseline blood sugar levels and monitor them throughout treatment. It is also a good idea to monitor lipid and cholesterol levels while taking these types of medications.

Extrapyramidal Symptoms (EPS)

People taking antipsychotic medications are at risk of developing side effects know as extrapyramidal symptoms. Extrapyramidal symptoms include parkinsonism, akathisia, dystonia, and tardive dyskinesia. These can include repetitive, involuntary muscle movements such as rocking a foot or lip smacking. People may have the urge to move constantly. It is very important to notify a doctor right away if any signs or symptoms of possible EPS are noticed.

Medications That Can Cause Extrapyramidal Symptoms This list is not all-inclusive	
Clozapine (Clozaril)	Haloperidol (Haldol)
Risperidone (Risperdal)	Chlorpromazine (Thorazine)
Olanzapine (Zyprexa)	Perphenazine (Trilafon)
Quetiapine (Seroquel)	Thioridazine (Mellaril)
Aripiprazole (Abilify)	Mesoridazine (Serentil)
Ziprasidone (Geodon)	Trifluoperazine (Stelazine)
Paliperidone (Invega)	Fluphenazine (Prolixin)

Treatment for Extrapyramidal Side Effects

If extrpyramidal side effects occur from neuroleptic medications, the physician may first try to reduce the dosage or switch to an atypical medication. Anti-parkinsonism drugs may be prescribed to help counter the effects of the neuroleptic medication. The most common ones used are Cogentin and Artane. Benzodiazepines have also been used to help alleviate symptoms. Anti-parkinsonism medications are not effective with tardive dyskinesia.

Abnormal Involuntary Movement Scales (AIMS)

If an individual supported is receiving any medication with tardive dyskinesia listed as a side effect, a nurse or doctor will need to complete an Involuntary Movement Scale (AIMS) on a monthly, quarterly, or every six month basis. The frequency in which the AIMS is completed is determined by the agency's accreditation and the funding source's requirements.

Extrapyramidal Side Effects

Side Effect	Sign / Symptom	When it Appears	What to do
Acute Dyskinesia	* Blinking * Writhing limbs * Protrusion of the tongue * Uncontrolled face movements	One to five days after starting the medication	Hold the medication and notify a nurse or doctor immediately
Akathesia	* Pacing * Foot rocking * Restlessness * Restless legs * Complaints of feeling "jittery" * "Crawling out of their skin"	Five days to two months after starting a new medication	Hold the medication and notify a nurse or doctor immediately
Dystonia	* Rigidity * Eyes rolled up * Locked jaw * Back arching * Head to one side or pulled back	One to five days after starting the medication	Hold the medication and notify a nurse or doctor immediately
Pseudo-Parkinsonism	* Tremor * Slow movement * Drooling * "Pill rolling" * Shuffling gait * Expressionless * Decreased blinking	Within five days to thirty days of starting a new medication	Hold the medication and notify a nurse or doctor immediately
Tardive Dyskinesia	* Chewing * Writhing limbs * Drooling * Blinking bursts * Protrusion of the tongue * Licking lips repeatedly * Uncontrolled face movements	Tardive means to "appear late" Occurs after a person has been on the med for over one year	Hold the medication and notify a nurse or doctor immediately

Practice Session

1. Which of the following is not considered a psychotropic medication?

    a. Anticoagulants

b. Antidepressants
   c. Antipsychotics
   d. Mod stabilizers

2. Which of the following is a common side effect of psychotropic medications?

   a. Increased thirst
   b. Drowsiness
   c. Photosensitivity
   d. All of the above

3. True or False. There may be an increased risk for people taking antipsychotic medications in developing diabetes.

4. True or False. If you suspect EPS symptoms in an individual supported, you should continue administering the medication as ordered and bring this to the attention of the nurse at his/her next visit.

Medication Administration Records

Medication Administration Records (MAR)

- A medication administration record is a legal document. It must be kept for a minimum of seven years, or according to your state's requirements.

- Blue or black is the only color of ink allowed on an MAR. If your agency has a policy or procedure in place stating that only black ink is allowed on an MAR, then only black ink is allowed on the MAR.

- If you make a mistake while documenting on an MAR, draw one single line through the mistake and write "error" along with your initials and the date. If needed, clarify on the back of the MAR.

- The medication administration record must contain:
    - The individual's name
    - Allergies
    - Time frame the MAR is being used from (example: Oct. 1, 2011 to Oct. 31, 2011)
    - Current Diet (If the person is on thickened liquid, medications must be administered using thickened liquids)
    - Primary Care Physician's Name and phone number
    - Additional information that may be required according to your state such as:
        - Date of Birth
        - Male or Female
        - Current Diagnoses

- Routine orders – These are medications that an individual takes on a daily basis.

- PRN medications – These are medications only taken if a person "needs" the medications. For example, if the individual is experiencing a headache and he/she has an order for motrin that may be given for pain, this medication would only be administered if the person experiences pain.

- Range orders are medications that are only administered for a certain period of time. For example, Amoxicillin 500mg P.O. TID x 7 days. This medication would only be administered for seven days and then discontinued.

Sample Medication Administration Record (MAR)

(Front Side of the MAR)

Medication	Schedule																					
Atenolol 50mg P.O. BID	8:00 AM	rl	rl	ll	ll	rl	lj	jl	jl	pl												
	8:00 PM	pl	pl	sl	lp	ip	ip	il	il	il												
Seroquel 50mg P.O. at Noon	Noon	pl	pl	sl	lp	ip	ip	il	il	il												
Lisinopril 10mg by mouth every evening																						
	5:00 PM	pl	pl	sl	lp	ip	ip	il	il	il												
Ambien 10mg P.O. at HS x 5 days																						
	8:00 PM	X	X	X	X	X	X	pl	il	pl		X	X	X	X	X	X	X	X	X	X	X
Tylenol 325mg by mouth every six hours prn pain or fever								pl														
	PRN																					

Charting For:	October 1, 2011 to October 31, 2011
Physician:	Dr. Williamson 573-555-5555
Diet:	Regular
Allergies:	Penicillin
Diagnosis:	Hypertension, Insomnia, Mood Disorder
Individual supported:	John Doe Date of Birth: 8-19-69 Sex: Male

(Back Side of MAR)

Name Signature		Initials	Name Signature		Initials
Robert Lenard	*Robert Leonard*	RL			
PRN Medication					

Date	Time	Medication	Reason Given	Initial	Time	Result	Initial
10-5-11	noon	tylenol 325mg PO	Headache	RP	1:00 PM	Asleep	RP

Review the sample MAR while reading the following information pertaining to MAR documentation.

- In order for a medication to be correctly written on a MAR, it must contain
    - Name of the medication
    - Dose (amount of the medication to be given)
    - Route (oral, eye medication, ear medication, suppository)
    - Frequency (how often is the medication to be given)
- After writing in the name of the medication, a time must be assigned for when to give the medication. This will be according to your agency's policy. Morning medications should always be assigned the first box in the schedule column, noon medications the second box, evening medications the third box and night time medications the fourth box. This will decrease the likelihood of medication errors because people will be looking for their specific time frame on the MAR when passing medications.
- When a medication is given, a person must initial the box corresponding to the time and date that the medication was given.
- Any person that writes on the MAR must print their name, sign their name and put their initials on the back of the MAR in the designated section. This allows people to compare the initials on the front of the MAR to the person's name on the back of the MAR so they know who administered a particular medication.
- The Ambien order on the sample MAR indicates that the medication is only being given for five days. Note how the days before and after the medication have an "X" and darker lines alerting staff to the specific time frame.
- All PRN medications must include the reason the medication is to be given. Example: Pain or fever. If the PRN order states for fever with no other indications, you would not be able to administer the medication for pain.
- Whenever a PRN medication is requested or felt needed:
    - Verify the medication has not been given on that particular day
    - If the medication has been given, verify that enough time has passed in order for you to administer the medication again (example: med ordered every 6 hours, as needed, for pain)
    - Once the medication has been administered, put your initials on the front of the MAR corresponding to the medication and date.
    - Complete the PRN section on the back of the MAR for every PRN administered. This includes listing the date and time the med was given, the medication given, the reason the medication was given, and the initials of the person administering the medication.
    - Approximately one hour after the PRN was administered; the second half on the back of the MAR needs to be completed. This includes the time, result of the PRN (For example: sleeping, pain level decreased, if non-verbal – no longer hitting his head, etc), and the

person who is checking the individual for the result will place their initials in the final box. This person does not have to be the same person that administered the medication.
- When rechecking the individual at approximately the hour mark, if the individual reports that the pain is worse or not improved, call your supervisor and/or nurse for further instructions. In the daily notes document everything that was done to help support the individual. These actions may include decreasing the noise level in the home, decreasing the light, placing a cool washcloth on their forehead, and/or asking them if they want to lay down.

Documenting a medication refusal

If an individual refuses a medication or medications, you simply have to put a circle in the box. You can leave the circle blank, you can put your initials inside the circle, or you can put an "R" for refuse inside the circle. You will also need to turn the MAR over onto the back and under the PRN documentation section, write out more details regarding the refusal. In addition to this, you will need to include in your daily notes everything you did to try and get the individual to take his/her medications before listing it as a refusal. Report all medication refusals according to your agency's policy. This may include contacting a supervisor, nurse, or doctor.

Atenolol 50mg P.O. BID	8:00 AM	rl	rl	ll	ll	rl	lj	jl	jl	pl	O								
	8:00 PM	pl	pl	sl	lp	ip	ip	il	il	il									

Date	Time	Medication	Reason Given	Initial	Time	Result	Initial
10-5-11	noon	tylenol 325mg PO	Headache	RP	1:00 PM	Asleep	RP
10-10-11	8:00 AM	Refused 8:00am medications		RP	n/a	n/a	n/a

Documenting when a person vomits a medication

If a person vomits a medication at any point after the medication has been administered, you cannot re-administer the medications. Once the medications have been swallowed, the pills start dissolving and breaking down. If a person vomits medications and you administer new medications, you are actually giving the person a higher dose than what was prescribed. Unless you have permission from a doctor or pharmacist, the medications cannot be re-administered.

If a medication is vomited, you simply put a circle in the box corresponding to the date and time. If you have already put your initials in the box indicating the medication was given, you circle your initials. You will also need to turn the medication administration record over and put an entry under the PRN section indicating that the medication was vomited.

Always report incidents where medications were vomited according to your agency's procedures. This may include contacting the supervisor, nurse, or doctor. Include additional information pertaining to the individual vomiting his/her medications in the daily notes.

Atenolol 50mg P.O. BID	8:00 AM	rl	rl	ll	ll	rl	lj	jl	jl	pl	O									
	8:00 PM	pl	pl	sl	lp	ip	ip	il	il	il	O									

PRN Medication							
Date	Time	Medication	Reason Given	Initial	Time	Result	Initial
10-5-11	noon	tylenol 325mg PO	Headache	RP	1:00 PM	Asleep	RP
10-10-11	8:00 AM	Refused 8:00am medications		RP	n/a	n/a	n/a
10-10-11	8:00 PM	Medication vomited 30 min after given		RP	n/a	n/a	n/a

Documenting a medication error (failure to administer)

If a staff person fails to administer a prescribed medication it is considered a medication error. In order to document this on the medication administration record, you will simply circle the box corresponding to the date and time the person did not receive the medication. You will also have to complete an entry on the back of the MAR giving additional information. An incident report will also need to be completed along with notifying the supervisor, nurse and doctor according to your agency's policy.

Lisinopril 10mg by mouth every evening																		
	5:00 PM	pl	pl	sl	lp	ip	ip	il	il	il	O							

PRN Medication							
Date	Time	Medication	Reason Given	Initial	Time	Result	Initial
10-5-11	noon	tylenol 325mg PO	Headache	RP	1:00 PM	Asleep	RP
10-10-11 5:00pm medications were not administered - incident report filed Robert Palmer							

Discontinuing a medication

If the doctor discontinues a medication, you will need to indicate the medication has been discontinued on the MAR, and also remove the medication so it is not at risk for accidentally being administered. Always follow your agencies procedure for discontinuing a medication. The following is an example of how it may be done.

Atenolol 50mg P.O. BID **Discontinued 10-10-11**	8:00 AM	rl	rl	ll	ll	rl	lj	jl	jl	pl	Medication Discontinued 10-10-11 RP
	8:00 PM	pl	pl	sl	lp	ip	ip	il	il	il	

On the last day that the medication was given, there needs to be a line drawn with a notation that the mediation has been discontinued along with the date and the initials of the person documenting it. There also needs to be a notation in the column with the medication orders.

New medication orders

If a new medication is ordered, you will need to fill the information in on the MAR and clearly indicate when the medication is to start. For example, the doctor started a new medication of Lasix 20mg P.O. QD to start the morning of 10-9-11.

Lasix 20mg By Mouth Every AM	8:00 AM			**START --->**												

Changes in current orders

If the doctor changes a dose of a medication that a person currently takes, the order must be discontinued and the new medication order written. It is not correct to simply cross out the old dose and write in the new dose.

For example, the Individual was taking Lasix 10mg by mouth daily, but the doctor decides to change it to 20mg by mouth daily.

Lasix 10mg By Mouth Every AM Medication Discontinued 10-10-11	8:00 AM	rp	rp	rp	rp	rp	rp	rp	rp	rp	Medication Discontinued 10-10-11 RP							

Lasix 20mg By Mouth Every AM	8:00 AM			START ------>											

For insurance reasons, the pharmacy may not be able to issue new 20 mg pills if you have quite a few 10mg tabs remaining. You will need to take the medication card back to the pharmacy and they will re-labe your medication card. The new card will state "Lasix 10mg, take two tabs by mouth daily".

The only person that can change a prescription label is a pharmacist. You cannot mark through the directions where it states take one tab and change it to two tabs. Again, only a pharmacist can change the directions on a prescription label.

Medication's ordered for every other day or monthly

Some medications may be ordered to be given either every other day (Q.O.D.) or on a weekly basis. It is very important that the medication administration record clearly indicates the dates the individual should receive the medication in order to decrease the risk of medication errors.

Warfarin 2.5mg by mouth every other day	8:00 AM	X	X	X	X	X	X	X	X	X	X	X	

Actonel 150mg P.O. Monthly		Noon	X	X	X	X	X	X	X	X	X	X	X	X	X	X	X	X	X	X	X	X

Medications requiring a pulse before being administered

Some medications require a pulse to be taken before the medication is administered. Digoxin and Propranolol are medications where this is required. In addition to documenting that the medication was administered, you also need to document the pulse. Parameters will be set by the ordering physician, but generally the medications are held for pulses below 60. If the medication is not administered, a circle will be placed in the box and documentation needs to occur on the back of the MAR. Always follow your agency's procedure for notifying the supervisor, nurse, or doctor when medications are held.

Propranolol 10mg by mouth every evening. Hold if Pulse less than 60.	Pulse	66	70	68	58							
	5:00 PM	RP	LJ	RP	O							

PRN Medication								
Date	Time	Medication	Reason Given	Initial	Time	Result	Initial
10-5-11	noon	tylenol 325mg PO	Headache	RP	1:00 PM	Asleep	RP
10-4-11 5pm Propranolol held - Pulse less than 60 (58) Robert Palmer							

Setting up and documenting medications for home visits

When consumers leave the ISL or group home for home visits, medications need to be sent with them. There are three approved methods.

Sending medications cards with guardians

Medication cards or bubble packs of medications may be sent home with the guardian if the guardian has been shown how to correctly use the bubble packs and agrees to this method of administering medications. If a person lives a long distance from the ISL or Group home you may want to consider setting up medications in a different manner in case medications are accidentally forgotten at the visit site and not returned. You may not be able to obtain additional medications from the pharmacy unless the medications are within a few days of being refilled.

When documenting on the MAR you will put a circle corresponding to the date and time each individual is not at the ISL or group home. It is recommended that you chart daily on the MAR and do not chart in advance that they are gone for a certain number of days. If the person comes back early, you will have to cross out each entry where the person came back early. On the back of the MAR you will need to make an entry for the days the person was not at the ISL, indicating the person was on a home visit.

Please note: Your agency may have specific codes that you need to use for home visits instead of a circle; this may include "H" for a home visit. This will be outlined in their policy and procedures. Regardless of the code, you are required to complete the back side of the MAR indicating who filled the medication planner.

Seroquel 50mg P.O. at Noon	Noon	pl	pl	sl	lp	ip	ip	il	il	il	O	O	O	O								

PRN Medication							
Date	Time	Medication	Reason Given	Initial	Time	Result	Initial
10-5-11	noon	tylenol 325mg PO	Headache	RP	1:00 PM	Asleep	RP
10-10-11	8:00 AM	Refused 8:00am medications		RP	n/a	n/a	n/a
10-10-11	8:00 PM	Medication vomited 30 min after given		RP	n/a	n/a	n/a
10-10-11 to 10-13-11 Individual supported was on a home visit with his guardian Robert Palmer							

Setting up medication pill planners

If having the guardian take the medication cards is not an option for a particular guardian, the medications may be set up using a medication planner. Medications are placed in the planner according to the orders written on the medication administration record. You must send information with the planner detailing the medications the individual will be taking at each medication time. The guardian needs to have this information so they know what medications they are giving and also as a resource in the event the individual has to be taken to urgent care or an emergency room.

In regards to documenting on the medication administration record, you will need to put a circle on the MAR corresponding to the date and time each medication was removed from the prescription bottle or medication card. You will also need to document on the back of the MAR indicating that the medications were removed and placed in a medication planner for a home visit.

Please note: Your agency may have specific codes that you need to use for home visits instead of a circle; this may include "H" for a home visit. This will be outlined in their policy and procedures. Regardless of the code, you are required to complete the back side of the MAR indicating who filled the planner.

Lisinopril 10mg by mouth every evening	5:00 PM	pl	pl	sl	lp	ip	ip	il	il	il	o	o	o	o	o	o				

PRN Medication							
Date	Time	Medication	Reason Given	Initial	Time	Result	Initial
10-5-11	noon	tylenol 325mg PO	Headache	RP	1:00 PM	Asleep	RP
10-10-11	8:00 AM	Refused 8:00am medications		RP	n/a	n/a	n/a
10-10-11	8:00 PM	Medication vomited 30 min after given		RP	n/a	n/a	n/a
10-10-11 to 10-15-11 Medication Planner filled for home visit with guardian Robert Palmer								

Please note that if the person comes back from a visit early, the only person allowed to administer the medications is the person that set up the medications in the planner. If the person comes home early and more than three days worth of medications would have to be wasted, you will need to work with your agency and pharmacy on how the medications may still be used. Exceptions may be made due to pharmacies being unable to issue additional medications. A pharmacist or RN may be able to identify and verify the medications based on markings on pills. As long as the pills are identified as the correct medication and the correct dose, your agency may give approval to administer the medications.

Setting up pill envelopes

Pill envelopes are also used in the same manner as setting up pills in a pill planner. Individual doses of the medication for each medication time are placed in a pill envelope. On the outside of the envelope you must list the Individual's name, date and time the medication is to be given, and the name and dose of each medication inside the envelope.

```
┌─────────────────────────────────────────┐
│            Pill Envelope                │
│                                         │
│  John Doe                               │
│  5:00 PM Medication     10-13-11        │
│                                         │
│  Lisinopril 10mg                        │
└─────────────────────────────────────────┘
```

Documentation on the MAR front and back is very similar to the method used when filling a medication planner.

| Lisinopril 10mg by mouth every evening | 5:00 PM | pl | pl | sl | lp | ip | ip | il | il | il | o | o | o | o | o | o | | | | | | | |

PRN Medication

Date	Time	Medication	Reason Given	Initial	Time	Result	Initial
10-5-11	noon	tylenol 325mg PO	Headache	RP	1:00 PM	Asleep	RP
10-10-11 to 10-15-11 Medication Envelopes filled for home visit with guardian				Robert Palmer			

Please note: It is illegal to repackage medications in any manner other than the methods described above. <u>Placing medications in a baggie or multiple different pills in a prescription bottle can lead to an arrest if discovered by law enforcement personnel.</u>

* Your pharmacist may be able to bubble pack medications for home visits or for school use.

Practice Session

1. Complete the medication administration record below:

    a. Show where the medication times of 8:00 am and 5:00 pm should be documented
    b. Show how to effectively start the medication on the 10th day of the month at 8:00am
    c. Put your initials on the MAR indicating you administered the 5:00 pm medication on the 13th day of the month

Atenolol 50mg by mouth at 8:00 am and 5:00 pm																																

2. Document that you administered Tylenol 325mg by mouth on October 4, 2011 at 4:00 pm. The individual was complaining of a headache and you had a valid order for this medication to be administered for pain. At approximately 5:00 pm the individual is observed sleeping.

Tylenol 325mg by mouth every six hours prn pain or fever	PRN																															

PRN Medication							
Date	Time	Medication	Reason Given	Initial	Time	Result	Initial

3. At 5:00 pm on November 05, 2011 the consumer refused his Folic Acid. Document the refusal on the medication administration record.

Folic Acid 1mg P.O. every evening	5:00 PM																		

PRN Medication							
Date	Time	Medication	Reason Given	Initial	Time	Result	Initial

4. The individual was seen by his doctor November 14th and the doctor discontinued his Celexa 40mg by mouth daily and ordered Celexa 20mg by mouth daily. Indicate the change on the MAR.

Celexa 40mg P.O. Daily	8:00 AM																		

5. The individual went to the doctor on November 10 and received an order for Amoxicillin 500mg by mouth three times daily for five days. The first dose of the medication will be given at bedtime. Set this medication order up on the medication administration record and make sure it is very clear to everyone when the individual should receive the medication.

Medication Administration

Administering Medications

(1) You need to be aware of the medications that you will be administering. Review side effect sheets for each medication, look them up in a drug reference book or obtain information from the nurse or pharmacist.
    - What is the medication used for (diabetes, hypertension, etc)
    - Know that the medication dose is within the normal range for the medication
    - Potential side effects

(2) Administer only medications that you have set up. You cannot administer any medication that another person set up. The person setting up medications is responsible from start to finish.

(3) Medications can only be administered within one hour of the scheduled time. If a medication is ordered for 8:00am, the medication can be given as early as 7:00am and as late as 9:00am. Anything outside of this time frame is considered a medication error unless approval is obtained by a pharmacist. If the medication was administered at 10:00 (outside of the time frame) and the individual received the same medication at noon, the person would be receiving the medications too close together.

(4) If your agency allows you to obtain telephone or verbal orders from a doctor, you will need to do the following:
    a. Listen carefully to the order given by the doctor.
    b. Write the order down on a piece of paper.
    c. Repeat the order back to the doctor and spell any words you are unsure of.
    d. After the order is correct, write Verbal order, the doctor's name, your name, date, and time.
    e. The order needs to be signed by the doctor within 7 days in order to be a legal order. This may be done by the doctor in person or by fax.
    f. The medication will need to be documented on the MAR and implemented as ordered.

(5) You cannot crush any medication without an order to crush it. If a person has a hard time swallowing, you need to bring this to the doctor's attention. The doctor will review all medications and determine if the medications may be crushed. Some medications may be enteric coated or time released and should not be crushed. The doctor may decide to change some medications to a liquid version. When medications are crushed it increases the rate at which they are absorbed.

(6) Enteric Coated medications are pills that have a special coating on them. They cannot be crushed or broken. The special coating allows them dissolve in the small intestine instead of the stomach.

(7) If a medication is scored (has an indentation down the center of the pill), it is the manufacturer's way of letting you know that it is safe to break the pill in half. If a pill is not scored you cannot break the pill in half unless you have permission from a pharmacist.

(8) If a person has difficulty taking their medications with liquids, it is okay to administer medications in pudding, applesauce, yogurt etc. You are not required to obtain a doctor's order to administer medications in this manner as long as the person is safe to eat these types of food **and** you are not hiding the medication in the food following a medication refusal.

(9) You must have an order for every medication that you are administering, including over the counter (OTC) medications. Some states allow you to use over the counter topical first aid ointments for minor cuts or bug bites without an order, but always verify this with your agency before using these medications.

(10) If you cannot read a doctor's handwriting, you must get clarification from the pharmacist or the doctor before administering the medication. When scripts are taken to the pharmacy, you will receive medication cards or prescription bottles with the medication name and directions typed out.

(11) Always check the expiration date on the medication container before administering the medication. Do not administer any medication that is out dated, not in the original or approved container, or any medication where the directions are missing, torn off, or illegible.

There are Seven Rights to Medication Administration

(1) **Right Person** – You have to make sure you have the right individual. If you have three individuals sharing one home, you need to make sure you are giving the medications to the right person.
   a. Most face sheets have photos of the individuals which can help identify the right person.
   b. Another staff person in the home should be able to help identify the correct individual.
   c. Ask the person to tell you their name. Never ask "Are you John Doe", as people have a tendency to say "yes" to everything. They may not have heard what you were saying and just said yes. If they have to verbally tell you their name that should eliminate the unintentional error.

(2) **Right Medication** – All medications have several names, brand names and generic names. You have to make sure you have the right medication before administering it. A lot of medications look alike and sound alike.

(3) **Right Dose** – You need to verify that you are giving the correct dose of the medication as ordered. If the doctor orders 500 mg of a specific medication and the pharmacy supplies 250 mg per pill, you will need to administer two tabs.

(4) **Right Route** – If the order is ophthalmic, if needs to be administered in the eye. Verify that you are familiar with the terminology for each route.

(5) **Right Time** – You have an hour before and after the scheduled time to administer each medication. Anything outside of this time frame is considered a medication error unless you have permission from a pharmacist or physician.

(6) **Right Dosage Form** – If the doctor prescribes the medication as a liquid, you can only administer the liquid form of this medication. The same is true of extended-release medications and other forms of medications. You have to follow the most current order received.

(7) **Right Charting** – In the court of law, if you did not document that you did something then you did not do it. You must document every medication administered and document all episodes for why a prescribed medication was not administered.

Step by Step Instructions for Administering Medications

(1) Concentrate on what you are doing. Remove yourself to a quiet area where you can concentrate on setting up each person's medication.
(2) Wash your hands.
(3) Take out one person's MAR and that person's signed physician's order sheet.
(4) Compare the MAR to the Physician order sheet to verify that you have current orders for every medication you will be administering.
(5) Take out only that individual's medications that you will be setting up. Everyone else's medications need to be kept in their storage area.
(6) Using the MAR start with the first medication listed and check to see if you need that medication for your current medication pass. (Example: 8:00 am medication).
(7) If you do need the first medication, verify that the medication card or prescription bottle and the MAR are both labeled with the correct individual's name. You also need to compare the medication name, dose, route, frequency, dosage form, and expiration date. Verify that the person is not allergic to the medication. If everything matches, set the card aside.
(8) Go down the list in order on the MAR and complete the same check for each medication. Only group together the medications needed for this particular medication pass. Once this has been completed, you have performed the first check. You are required to do **three** checks before administering any medication.
(9) On the second check, you will go back through the medications you have grouped together. You will verify that the medication is correct, again checking the dose, route, frequency, and dosage form. You will also verify that the card and prescription label are both labeled with the correct individual's name, and that the person is not allergic to the medication.
(10) As you finish checking each medication with the second check, you will remove the pill(s) from the bubble card or prescription bottle. Hold a medication cup to the back of the bubble you will be "popping" and push the medication into the cup. Your hands should not touch the medication. If administering from a prescription bottle, pour the medication into the cap of the bottle, and then from the cap place the medication into the medication cup.
(11) If removing the medication from a medication card, it is a good idea to put your initials, med time, and date next to the empty bubble where the medication was removed. This gives additional documentation that the medication was administered in the event the staff person forgets to initial the MAR for the medications administered.
(12) Continue in order down the MAR, placing the medications into the medication cup after performing the check for each medication. You now have completed the second required check.
(13) You must go through the medications one last time. You need to go in order according to the MAR and verify that the prescription label, medication, dose, route, frequency and dosage form match. You also need to check one last time that the person is not allergic to the medication and that you are administering from the correct person's medication card or bottle. Once this is completed the third check has been done. The medications need to be secured and not left out before taking the medication(s) to the individual.

(14) Take the medications to the individual or have the individual come to you. Verify that you have the correct person. Once you are positive that you have the right person, you may administer the medications.
(15) Stay with the individual until all of the medications have been taken or applied. Do a mouth check if needed, as some individuals may pocket medications or fail to swallow them.
(16) Return to the MAR and document that the medications were administered. You cannot document the medications **until** the medications have been swallowed or applied.
(17) Make sure your signature, printed name and initials are on the back of the MAR. If PRN medications were administered, or if the individual did not receive a prescribed medication, make sure this documentation is completed under the PRN section on the back of the MAR.

Additional tips:

You may want to go through the MAR and count to see how many pills you should have in your medication cup and compare it to how many pills you actually have in your cup.

As you pull up each medication you may want to put a small dot in the box that you will be initialing indicating that the medication was administered. If you are called away while setting up your medications, you will know exactly where you left off. It also makes it easier to come back and sign for the medications after the individual takes his/her medications.

Steps (Abbreviated format):

1. Concentrate
2. Wash your hands
3. Compare the MAR to Physician Order Sheet
4. First Check – Select only meds needed for that med pass, compare prescription label and MAR for individual's name, allergies, med, dose, route, frequency, dosage form.
5. Second Check – Recheck, compare MAR and prescription label to individual's name, allergies, med, dose, route, frequency, dosage form. With this check you also put the med in the med cup. Initial, date, and time the medication card where the med was removed.
6. Third check – Recheck the MAR to the prescription label before putting the medication back in storage AND before administering the medication to the individual. Compare the MAR and prescription label to the individual's name, allergies, med, dose, route, frequency, and dosage form.
7. Administer the medication, verify you have the right person, stay with the individual until med is taken.
8. Initial the MAR indicating the medication was administered.
9. Make sure your name, signature, and initials are on the back of the MAR.
10. Complete the PRN section for any PRN medications administered, refusals, etc.

Self Administer or Partial Participation Medication Administration

Some individuals are very responsible and want to administer their own medications. The treatment team will need to meet and discuss if this is a possibility for this particular individual. In order for this to occur the case manager, agency representative, guardian, and physician must agree that the individual is responsible enough to learn how to self administer medications or participate in a partial participation program.

If everyone has agreed to move forward with self administering medications, the physician will need to write an order and the individualized plan will need to be updated. The individualized plan will describe the needed steps needed in order for the individual to learn how to administer his/her own medications.

(1) There needs to be education documented that the individual knows the names of the medication he is taking and the reason for each medication.
(2) He needs to be shown how to properly read the prescription label and administer the medications as ordered.
(3) He will need to demonstrate how to self administer his medications on several occasions with an experienced staff person watching to verify that he is doing it correctly. This training will continue until the agency's representative is satisfied that the individual can safely take medications without assistance.
(4) The medications need to be checked on a routine basis to verify that the medications are being taken correctly. This may include counting the medications compared to the number of doses that should be missing on a particular day. The frequency of this monitoring needs to be documented in the individualized plan.
(5) Reordering of medications may be done by the agency or the individual. Either way, the steps for reordering medications are part of the individualized plan.

Some people may not be ready to completely self administer their own medications. If the person is seeking independence the treatment team may want to consider a partial participation program. In order to meet the requirements, the case manager, agency representative, guardian, and doctor have to agree with the plan.

Once everyone agrees, the doctor will need to write an order allowing the partial participation medication administration. The individualized plan will need to be updated detailing the steps of the partial participation of self medication.

(1) Provide education and document that the individual knows the names of the medication he is taking and the reason for each medication.
(2) He needs to assist in filling a medication planner for a specified number of days. By assisting, the individual should be reading the labels, removing the pills from their package and placing the pills in the correct slot in the planner. The staff person watches and offers assistance as needed. <u>The staff person is responsible in making sure that the planner is filled correctly</u>.
(3) The back of the MAR will be completed by the staff person to indicate that the planner was filled for the specified number of days per the partial participation program.

(4) When it is time to take the medications, the individual must initiate getting his own medications without prompts. He will be the one to remove the pills from the planner, obtain his own drink, and swallow his pills.

(5) Staff will need to check the planner on a routine basis as outlined in the individualized plan. This is to verify that all of the medications have been taken according to how they are prescribed.

- With either of the two methods listed above, if it is determined that the individual is not administering medications correctly, additional training will need to occur or it will be determined that this type of administration is not safe and appropriate.

- The individualized plan may contain additional steps as required. Always follow your state's requirements for self or partial participation rules.

Administering Ophthalmic (Eye) Medications

1. Verify you have a current order for the medication (copy of script or physician order sheet).
2. Compare the Medication Administration Record to the prescription label three (3) times. Verify the following:
    a. Right Person
    b. Right Dose
    c. Right Route
    d. Right Med
    e. Right Time
    f. Right dosage form (drops or ointment)
    g. Verify no allergy to this medication
    h. Check expiration date on the container
3. Wash your hands and put on gloves.
4. Take the eye medication to the individual along with a tissue, and explain the procedure.
5. Instruct the individual to tilt their head back.
6. Gently pull down on the lower eyelid and place the number of drops or strip of ointment where you have pulled the eyelid down. Do not place the medication directly on the eye. Do not let the bottle or tube touch the person's eye or eyelid. Make sure you are placing the medication in the correct eye as ordered.
7. If applying eye drops, immediately place a gloved finger over the tear duct and apply gentle pressure for approximately 3 seconds.
8. If administering eye drops to the opposite eye, gently pull down the lower eyelid and place the number of drops or strip of ointment where you have pulled the eyelid down. If you are applying drops, apply gentle pressure to the tear duct using a different gloved finger.
9. Instruct the individual to blink several times to spread the medication around the eyes. Use the tissue to wipe away any medication that may be on the outside of the person's eye.
10. Return the bottle to storage.
11. Remove your gloves and wash your hands.
12. Document the administration of the eye medication on the medication administration record.

Administering Otic (Ear) Drops

1. Verify you have a current order for the medication (copy of script or physician order sheet).
2. Compare the Medication Administration Record to the prescription label three (3) times. Verify the following:
    a. Right person
    b. Right Dose
    c. Right Route
    d. Right Med
    e. Right Time
    f. Verify no allergy to this medication
    g. Check expiration date on the bottle
3. Wash your hands and put on gloves.
4. Take the medication to the individual along with a tissue and explain the procedure.
5. Instruct the individual to lay down or tilt their head with the correct ear facing up.
6. Gently pull the mid-outer ear back and up in order to open up the ear canal.
7. Administer the number of drops ordered by the physician. To prevent contamination, do not touch any part of the bottle to the inner ear.
8. The individual needs to stay in this position for at least three minutes.
9. If both ears require ear drops, after three minutes ask the person to turn to the opposite side and complete the steps again.
10. Return the bottle to storage.
11. Remove gloves and wash your hands.
12. Document the administration of the ear medication on the medication administration record.

Administering Nasal (Nose) Medications

1. Verify you have a current order for the medication (copy of script or physician order sheet).
2. Compare the Medication Administration Record to the prescription label three (3) times. Verify the following:
    a. Right Person
    b. Right Dose
    c. Right Route
    d. Right Med
    e. Right Time
    f. Verify no allergy to this medication
    g. Check expiration date on the bottle
3. Wash your hands and put on gloves.
4. Take the medication to the individual along with a tissue and explain the procedure.
5. Apply the medication according to the physician's order. (ie: Two sprays both nostrils). In order to do this correctly, the bottle will be placed slightly inside the nose.
6. Instruct the individual not to blow his/her nose for at least 15 minutes after instilling the medication.
7. Clean the top of the nasal container.
8. Return the bottle to storage.
9. Remove your gloves and wash your hands.
10. Document the administration of the nasal medication on the medication administration record.

Administration of Inhaled Medications

1. Verify you have a current order for the medication (copy of script or physician order sheet).
2. Compare the Medication Administration Record to the prescription label three (3) times. Verify the following:
    a. Right person
    b. Right Dose
    c. Right Route
    d. Right Med
    e. Right Time
    f. Verify no allergy to this medication
    g. Check expiration date on the container
3. Wash your hands and put on gloves.
4. Take the medication to the individual and explain the procedure.
5. For inhalers with metal canisters, shake thoroughly.
6. Remove the protective cap.
7. With the inhaler close to the person's mouth, have them exhale as much as they can. Before inhaling instruct them to enclose their lips around the mouth piece while the canister is held vertically.
8. Have the individual inhale slowly through the mouth piece while pressing firmly on the canister one time.
9. Remove the canister and instruct the person to hold their breath for at least 3 seconds.
10. If more than one puff is ordered, repeat the steps beginning with step #7.
11. If a second inhaler (different from the inhaler just administered) is scheduled at the same time, you must wait at least 5 minutes before administering the second inhaler. Follow the same steps as described above.
12. If the inhaler administered is a steroid inhaler, the individual needs to rinse their mouth out with water after using the inhaler.
13. Clean the mouth piece and replace the protective cap on the inhaler.
14. Return the inhaler to storage.
15. Remove gloves and wash hands.
16. Document the administration of the inhaler on the medication administration record.

Spacers:

If a person has a hard time following the directions, you may obtain an order for a spacer. This is a tube like device that fits on the end of the inhaler. You administer the inhaler in the same way following steps 7 and 8, but after instead of removing the inhaler and asking them hold their breath, they continue breathing in and out through the spacer for approximately 5 breaths.

Nebulizers:

If a person cannot complete the steps for the inhaler or the inhaler with a spacer, the person may need an order for a nebulizer machine. This will require a nurse delegation

Administering Topical Medications

1. Verify you have a current order for the medication (copy of script or physician order sheet).
2. Compare the Medication Administration Record to the prescription label three (3) times. Verify the following:
    a. Right person
    b. Right Dose
    c. Right Route
    d. Right Med
    e. Right Time
    f. Verify no allergy to this medication
    g. Check expiration date on the container
3. Wash your hands and put on gloves.
4. Take the medication to the individual and explain the procedure.
5. Examine the affected area. If ordered, cleans the area with soap and water, or as instructed.
6. Apply medication according to the instructions.
7. Never let the container touch the wound. If using gauze, you will need a new piece of gauze each time you need to get additional medication from the container to avoid contamination.
8. Return the container to storage.
9. Remove the gloves and wash your hands.
10. Document the administration of the topical medication on the medication administration record.

Administering Vaginal Medications

1. Verify you have a current order for the medication (copy of script or physician order sheet).
2. Compare the Medication Administration Record to the prescription label three (3) times. Verify the following:
    a. Right Person
    b. Right Dose
    c. Right Route
    d. Right Med
    e. Right Time
    f. Right Dosage Form
    g. Verify no allergy to this medication
    h. Check expiration date on the bottle
3. Wash your hands and put on gloves.
4. Take the medication to the individual along with a tissue and explain the procedure.
5. Provide privacy and position the individual on her back with knees bent and legs separated.
6. Use applicator and lubrication if indicated.
7. Spread the labia with one hand and gently insert the applicator or medication into the vagina with the other hand. Angle the applicator down toward the tailbone. It should be inserted approximately 2 inches. Do not force it.
8. If using an applicator, push the plunger while holding the applicator still.
9. Remove the applicator and instruct the individual to remain still for 30 minutes.
10. Remove gloves and wash hands.
11. Document the administration of the vaginal medication on the medication administration record.

Administering Rectal Suppositories

1. Verify you have a current order for the medication (copy of script or physician order sheet).
2. Compare the Medication Administration Record to the prescription label three (3) times. Verify the following:
    a. Right person
    b. Right Dose
    c. Right Route
    d. Right Med
    e. Right Time
    f. Verify no allergy to this medication
    g. Check expiration date on the bottle
3. Wash your hands and put on gloves.
4. Take the medication to the individual along with a tissue and explain the procedure.
5. Provide privacy.
6. Have the person laying on their on their left side or on their hands and knees with their bottom in the air.
7. Unwrap suppository and lubricate it along with your finger.
8. Lift the upper buttock to expose the rectum.
9. Slowly insert the suppository past the sphincter.
10. Encourage the individual to continue lying down for approximately 15 minutes.
11. Remove gloves and wash hands.
12. Document the administration of the rectal medication on the medication administration record.

Administering Prefilled Enema bottles

1. Verify you have a current order for the medication (copy of script or physician order sheet).
2. Compare the Medication Administration Record to the prescription label three (3) times. Verify the following:
    a. Right person
    b. Right Dose
    c. Right Route
    d. Right Med
    e. Right Time
    f. Verify no allergy to this medication
    g. Check expiration date on the bottle
3. Wash your hands and put on gloves.
4. Take the medication to the individual and explain the procedure.
5. Provide privacy.
6. Have the individual lay on his/her left side or on their knees with their bottom in the air.
7. Remove the cap from the enema, the end is already pre-lubricated.
8. Insert the tip of the enema approximately 2 inches.
9. Slowly squeeze all of the liquid into the rectum. Do not squeeze too fast; this can lead to cramping and discomfort.
10. Pull the bottle out and instruct the individual to hold the liquid inside of them until they feel like they cannot hold it any longer. The longer they can hold the enema, the better results will be achieved.
11. When they cannot hold the enema any longer, instruct them to go to the bathroom and release it into the toilet.
12. You may need to assist the person in cleaning up following the procedure.
13. Remove your gloves and wash your hands.
14. Document the administration of the enema on the medication administration record. Always document the results. If this is a PRN medication, complete the back side of the MAR including the result section. If it is not a PRN medication, document the results in the daily notes.

Note:

If the person is low functioning or does not have proper muscle control (quadriplegics), you may have to administer the medication in a bed that is lined with disposable chuxs. Administer the medication as described above, but the results will be quicker if the person can't hold in the fluid. You will have to clean the person up while they remain in bed.

Administering Oxygen

1. Verify you have a current order for the medication (copy of script or physician order sheet).
2. Compare the Medication Administration Record to the prescription label three (3) times. Verify the following:
    a. Right Person
    b. Right Dose (Liters per minute)
    c. Right Time
    d. Right Method (Oxygen concentrator or Oxygen tank)
3. Wash your hands.
4. Take the oxygen and either mask or nasal cannula to the individual and explain the procedure.
5. Hook the tubing of the mask or cannula to the tank or concentrator.
6. If the oxygen machine has a humidifier chamber, make sure that distilled water is filled to the appropriate water level.
7. Turn the machine on and verify the setting matches the order (ie: 3 liters per min).
8. Place the mask or nasal cannula on the person according to the instructions.
9. Wash your hands.
10. Document the administration of Oxygen on the medication administration record.

Note:

Oxygen is flammable. The person needs to stay away from open flames such as a gas stove, fireplace, burning candles, incense, and even smoking while hooked up to oxygen.

The front door of the residence must have a sign stating "No Smoking, Oxygen In Use".

Clean equipment as ordered. Do not use vasoline on nasal cannula as it is contraindicated.

Mask, nasal cannula and tubing all needs to be replaced as ordered.

Administering Transdermal Medications

1. Verify you have a current order for the medication (copy of script or physician order sheet).
2. Compare the Medication Administration Record to the prescription label three (3) times. Verify the following:
    a. Right Person
    b. Right Dose
    c. Right Route
    d. Right Med
    e. Right Time
    f. Verify no allergy to this medication
    g. Check expiration date on the box or patch itself
3. On top of the patch record your initials, date and time.
4. Read on the package insert where the patch can be applied.
5. Wash your hands and put on gloves.
6. Take the medication to the individual and explain the procedure.
7. Verify that the old patch has been removed.
8. Clean the skin with mild soap and water or as directed, the skin must be dry before the patch is applied.
9. Apply the patch to healthy skin that is in an area approved by the manufacturer (abdomen or shoulder blade region). The patch cannot be over cuts, sores, bruises, etc. If the person has a lot of hair in that region, the person will need to be shaved in order for the patch to make good skin contact.
10. The patch must be rotated to a new location each time a new patch is applied.
11. Make sure the patch is secure.
12. Discard wrappers and the old patch if it was removed at this time. Discard according to your agency's policy.
13. Take off gloves and wash your hands.
14. Document the administration of the transdermal patch on the medication administration record. Note the location where the patch was placed. The next person will use this information so they know where on the body the patch is located in order to remove the patch. You also need to know where the last patch was applied so you can verify the old patch was removed before applying a new patch.

Note:

If a patch comes off before it is scheduled to be removed, you will need to use medical tape and re-secure the patch. You cannot replace the patch with a new patch without a doctor's order.

Special Instructions for Administering Medications

Sublingual and Buccal Medications:

- Sublingual medications are placed under the tongue and allowed to dissolve without swallowing.
- Buccal mediations are placed just inside the cheek of the person's mouth and allowed to dissolve without swallowing.
- Do not give any liquid with these medications, they are designed to dissolve in the mouth and be absorbed directly into the blood stream from the mouth.
- Sublingual or buccal medications may be prescribed for chest pain, during a seizure, or to help control someone's behavior. If you administer a medication in this manner, the medication goes straight into the blood stream and you get fairly fast results. If you were to swallow the medication the result of the medication would be delayed.

Measuring liquids:

- Always shake liquid medications unless the bottle says not to shake it. By shaking the medication you are making sure that the active and inactive ingredients are well mixed. Certain medications may have warnings for you not to shake (eye drops) due to the risk of putting air bubbles into the solution. If a medication says not to shake, you will need to gently roll the medication in your hands.
- Always place your palm against the label which contains the directions when pouring. If medication spills down the bottle, this will prevent the directions from becoming illegible.
- Medications need to be poured with the measuring cup on a flat surface and at eye level in order to verify that you have poured the correct dose.
- Use a measuring device that is calibrated. Do not use silverware used at a dinner table. Examples of appropriate devices include calibrated measuring cups or a syringe.

Nursing Delegation

- Delegation is when a nurse teaches a non-licensed staff person how to perform a particular nursing activity or task.
- The nurse will assess the individual needing the delegated nursing task to determine if his/her condition is stable and that health-related delegation is appropriate.
- The nurse will complete a statement of delegation for each individual who requires delegation.
- The delegated employees will receive individual specific training detailing the task and the steps required in completing the task. The nurse will also review what needs to be reported back to the nurse in regards to the specific task.
- The nurse will observe the skills of the personnel at least annually or more often as needed.
- The nurse has the authority to withdraw the delegation at any point if the nurse feels that the task is not or will not be completed safely.
- The nurse will provide on-going supervision of the delegated nursing activity and will comply with all rules and regulations set forth by the State Board of Nursing for the state that the nurse is operating in.
- Agencies must have a procedure in place for documenting and tracking all nursing delegations.
- Always consult with your State Board of Nursing to determine which tasks may or may not be delegated.

Examples of Nursing Delegation Tasks
Foley Care
Colostomy Care
Administering Glucagon
Suctioning
Tube Feedings
Nebulizer Machines
C-Pap Machines
Epi Pen Administration
Mace Tube Treatments

Practice Session

1. True or False. There are seven rights to medication administration.

2. True or False. If a person has a hard time swallowing, you may crush their medication for ease in swallowing.

3. True or False. If a medication is over-the-counter, you may administer the medication without an order from a doctor to an individual supported.

4. True or False. You do not need to wear gloves when administering eye medications.

5. A physician's order reads: Amoxicillin 500mg T.I.D. by mouth. The pharmacy supplied Amoxicillin 250mg. You would administer:

    a. 1 capsule twice daily
    b. 2 capsules twice daily
    c. 1 capsule three times daily
    d. 2 capsules three times daily

6. Which of the following medication administration techniques are you not qualified to perform after completing this course.

    a. Insulin injections
    b. Rectal medications
    c. Inhalers
    d. Ophthalmic medications

Specialized G/J Tube Training

- The purpose of feeding tubes is to ensure normal growth and development or to maintain health and wellness.

- People are candidates for tube feeding when they cannot meet their nutritional requirements.

- Gastrostomy (G-Tube). This is a flexible tube or button placed into the stomach through an opening in the abdominal wall. This tube provides feeding or medication administration along with venting and drainage.

- Jejunsotomy (J-Tube). This is placed in the jejunum (middle section of the small bowel) through an opening in the abdominal wall or through a specific port on a G-Tube. This tube provides food and liquids directly into the jejunum.

- People that receive no food or liquid orally need to have good mouth care performed throughout the day. It is a good idea to brush teeth after each meal, lubricate lips as needed and receive routine dental care as directed.

- The primary care physician needs to be aware of every medication prescribed to the individual. Medications such as enteric coated and extended release should be avoided. If medications are available in a liquid form, this would be preferred over having to dissolve medications in water. The doctor will be the one to make the decision on the appropriate medication choices.

- Be proactive. If you are having a particular problem with a medication not dissolving well and clogging the tube, talk with the doctor, he may be able to prescribe an alternative medication that may serve the same purpose.

- People that receive tube feedings only cannot eat or drink anything by mouth. The amount of water prescribed to the person by the physician is the only amount of water that may be administered. The amount of water is calculated (generally by a dietician) to make sure the person is receiving enough water. Too much water can thin out electrolytes and cause additional health problems. <u>YOU MAY ONLY GIVE THE AMOUNT OF WATER ORDERED.</u>

- Always give water before <u>AND</u> after feedings and medication administration. This verifies the tube is open and working correctly before medications or the feeding is administered and also clears the tubing of the feeding or medications when finished.

Example:

John Doe
Delegation of Bolus Tube Feeding

Duties delegated by _____, RN

Duties to be performed at _____ House by staff. On-going training will be done as needed.

Procedure:
John receives scheduled tube feedings daily. This is his only method to receive nutrition, medication and fluids. He is not to have any food, drink or medication by mouth. To ensure John's health, safety and welfare, it is important that this procedure be followed consistently by all staff working with him. If at any time you have questions, concerns or comments, please contact Health Services as soon as possible.

1. Gather needed supplies.
    A. amount of product ordered for the feeding
    B. amount of water ordered
    C. syringe

Tube feeding solution and water should be room temperature to luke warm, not hot or cold. Extreme temperatures can cause injury and/or cramping.

After each use wash the syringe with mild dish detergent. Additionally, syringe should be put in the dishwasher every Monday.

2. John should be in a position that is comfortable to him, but he has to be at least 45 degrees in elevation. This can be in either a wheelchair, elevated in bed, or a recliner.

3. Wash your hands and apply disposable gloves.

4. Verify that you have the correctly ordered product for the individual's tube feeding. Check the expiration date and wipe off the top of the can.

5. Check for residual by removing the cap on the feeding tube and inserting a syringe with the plunger. Pull back on the plunger and the stomach's content should enter into the syringe. If you are able to obtain _____ cc of fluid hold the feeding and notify a nurse or doctor. (Generally 60cc or more requires you to hold the feeding, but always follow the doctor's instructions).

6. Check for placement by using a stethoscope and placing this over the person abdomen approximately one inch away from the belly button. When you administer the water that is ordered prior to the feeding you should be able to hear the 'gurgling' of the water enter the stomach. If you cannot hear this sound, do not administer the feeding and notify a nurse or doctor.

7. Bolus Tube Feeding Administration:

    a. Remove the cap from the feeding tube
    b. Insert the syringe immediately into the opening
    c. Add the amount of water ordered prior to the feeding
    d. Shake the tube feeding solution and then open the can. Pour the liquid directly into the syringe.

- e. Hold the syringe 8 to 10 inches above the person to allow the liquid to enter by gravity.
- f. Do not force the liquid into the tube
- g. Add the amount of water ordered following the feeding
- h. Remove the syringe and immediately replace the cap
- i. Avoid letting air get into the stomach – this can cause discomfort

8. The person needs to remain upright for at least one hour after the procedure.

9. Wash/clean any spilled liquid off the tube or stomach with mild soap and water. Keep the area clean and dry. Observe for redness, swelling, warmth, drainage, bleeding, unusual color or odor. Report these findings to a nurse or doctor immediately.

10. Clean equipment. It should be cleaned with a mild dish detergent and a brush in the sink. Air Dry.

11. Remove gloves and wash hands.

12. Document the tube feeding.

STOP THE TUBE FEEDING IF THE PERSON EXPERIENCES:

- NAUSEA OR VOMITING
- COUGHING
- CHOKING
- DIFFICULTY BREATHING
- ABDOMINAL OR CHEST PAIN
- FORMULA OBSERVED IN THE MOUTH

OBSERVE AND REPORT IMMEDIATELY THE FOLLOWING POTENTIAL COMPLICATIONS:

- Abdominal distention
- Discomfort/fullness/gas
- Constipation
- Diarrhea
- Fever
- Wheezing, gurgling, whistling
- Change in behavior

Example:

John Doe
Delegation of Continuous Tube Feeding

Duties delegated by _____, RN

Duties to be performed at _____ House by staff. On-going training will be done as needed.

Procedure:
John receives scheduled tube feedings daily. This is his only method to receive nutrition, medication and fluids. He is not to have any food, drink or medication by mouth. To ensure John's health, safety and welfare, it is important that this procedure be followed consistently by all staff working with him. If at any time you have questions, concerns or comments, please contact Health Services as soon as possible.

1. Gather needed supplies.
    - A. amount of product ordered for the feeding
    - B. amount of water ordered
    - C. Tube feeding bag (not to be reused)
    - D. Pole to hang the feeding bag or a feeding pump

Tube feeding solution and water should be room temperature to luke warm, not hot or cold. Extreme temperatures can cause injury and/or cramping.

2. John should be in a position that is comfortable to him, but he has to be at least 45 degrees in elevation. This can be in either a wheelchair, elevated in bed, or a recliner.

3. Wash your hands and apply disposable gloves.

4. Verify that you have the correctly ordered product for the individual's tube feeding. Check the expiration date and wipe off the top of the can.

5. Check for residual by removing the cap on the feeding tube and inserting a syringe with the plunger. Pull back on the plunger and the stomach's content should enter into the syringe. If you are able to obtain _____ cc of fluid hold the feeding and notify a nurse or doctor. (Generally 60cc or more requires you to hold the feeding, but always follow the doctor's instructions).

6. Check for placement by using a stethoscope and placing this over the person abdomen approximately one inch away from the belly button. When you administer the water that is ordered prior to the feeding you should be able to hear the 'gurgling' of the water enter the stomach. If you cannot hear this sound, do not administer the feeding and notify a nurse or doctor.

7. Continuous Tube Feeding:

    a. Clamp off the tubing connected to the feeding bag.
    b. Add the amount of water ordered to the bag.
    c. Prime the tubing (get all the air out – water should be in the tubing all the way to the end). Unclamp the tubing until the water gets to the end of the tube and then re-clamp it.
    d. Insert the tubing into the uncapped G/J tube. Unclamp the tubing and let the water enter into the

stomach. Clamp off the tubing before all off the water enters the tubing to avoid air from entering the stomach.
   e. Shake the tube feeding solution and then open the can(s). Pour the liquid directly into the feeding bag.
   f. If using gravity, hang the bag on a pole and adjust the number of drops per minute using the roller attached to the clamp. Pushing the roller up should slow down the drops and pushing the roller down should speed up the drops per minute. The nurse should calculate the drops per minute for you.
   g. If you are administering the feeding with a feeding pump, insert the tubing through into the machine according to the manufacturer's directions. Make sure the machine is set correctly (example: 60cc/hour). Turn the machine on.
   h. When the feeding has ended, add the amount of water ordered and either let this run through the machine or by gravity (according to your orders). Avoid letting air get into the stomach – this can cause discomfort.
   i. Remove the tubing from the G/J tube and immediately replace the cap.

8. The person needs to remain upright for at least one hour after the procedure.

9. Wash/clean any spilled liquid off the tube or stomach with mild soap and water. Keep the area clean and dry. Observe for redness, swelling, warmth, drainage, bleeding, unusual color or odor. Report these findings to a nurse or doctor immediately.

10. Remove gloves and wash hands.

11. Document the tube feeding.

STOP THE TUBE FEEDING IF THE PERSON EXPERIENCES:

- NAUSEA OR VOMITING
- COUGHING
- CHOKING
- DIFFICULTY BREATHING
- ABDOMINAL OR CHEST PAIN
- FORMULA OBSERVED IN THE MOUTH

OBSERVE AND REPORT IMMEDIATELY THE FOLLOWING POTENTIAL COMPLICATIONS:

* Abdominal distention
* Discomfort/fullness/gas
* Constipation
* Diarrhea

* Fever
* Wheezing, gurgling, whistling
* Change in behavior

Example:

John Doe
Delegation of Medication Administration Per Tube

Duties delegated by _____, RN

Duties to be performed at _____ House by staff. On-going training will be done as needed.

Procedure:
John receives scheduled tube feedings daily. This is his only method to receive nutrition, medication and fluids. He is not to have any food, drink or medication by mouth. To ensure John's health, safety and welfare, it is important that this procedure be followed consistently by all staff working with him. If at any time you have questions, concerns or comments, please contact Health Services as soon as possible.

1. Gather needed supplies.
   A. amount of water ordered
   B. syringe
   C. medication

Tube feeding solution and water should be room temperature to luke warm, not hot or cold. Extreme temperatures can cause injury and/or cramping.

After each use wash the syringe with mild dish detergent. Additionally, syringe should be put in the dishwasher every Monday.

2. John should be in a position that is comfortable to him, but he has to be at least 45 degrees in elevation. This can be in either a wheelchair, elevated in bed, or a recliner.

3. Wash your hands and apply disposable gloves.

4. Crush any pills or empty any capsules. They must be crushed to a powdery consistency. Add warm water to the medication cup. Medications must dissolve or they can clog the tube. Medications need to be prepared separately. Always mix medications with water and not the feeding. Always verify you have the Correct Person, Medication, Dose, Route, Time, and Dosage Form. Also verify that the person is not allergic to the medication and the medication is not expired.

5. Check for placement by using a stethoscope and placing this over the person abdomen approximately one inch away from the belly button. When you administer the water that is ordered prior to the medications, you should be able to hear the 'gurgling' of the water enter the stomach. If you cannot hear this sound, do not administer the feeding and notify a nurse or doctor.

6. Medication Administration:
   a. Remove the cap from the feeding tube
   b. Insert the syringe immediately into the opening
   c. Add the amount of water ordered prior to medications
   d. Immediately add the dissolved medications in water to the syringe, Add the prescribed amount of water between each medication.
   e. If liquid medications are ordered they should be administered before medications that are

dissolved in water. Always give a small amount of water between medications as prescribed.
- f. Medications should be delivered slow and steadily, hold the syringe 8 to 10 inches above the person to allow the liquids to enter by gravity
- g. Do not force medication or fluid into the tube
- h. Add the amount of water ordered following medication administration
- i. Remove the syringe and immediately replace the cap

7. The person needs to remain upright for at least one hour after the procedure.

8. Wash/clean any spilled liquid off the tube or stomach with mild soap and water. Keep the area clean and dry. Observe for redness, swelling, warmth, drainage, bleeding, unusual color or odor. Report these findings to a nurse or doctor immediately.

9. Clean equipment and pill crusher. It should be cleaned with a mild dish detergent and a brush in the sink. Air Dry.

10. Remove gloves and wash hands.

11. Document the medication administration per feeding tube.

STOP THE ADMINISTRATION OF MEDICATIONS IF THE PERSON EXPERIENCES:

- NAUSEA OR VOMITING
- COUGHING
- CHOKING
- DIFFICULTY BREATHING
- ABDOMINAL OR CHEST PAIN
- FORMULA OBSERVED IN THE MOUTH

OBSERVE AND REPORT IMMEDIATELY THE FOLLOWING POTENTIAL COMPLICATIONS:

- Abdominal distention
- Discomfort/fullness/gas
- Constipation
- Diarrhea
- Fever
- Wheezing, gurgling, whistling
- Change in behavior

Specialized Diets

Specialized Diets

Thickened Liquids:

Thickened liquids are ordered for people who are at risk for aspirating thin liquids when drinking. Aspiration is when a person drinks and small amounts of the fluid enters the person's lungs. This may cause the person to cough as the body tries to get rid of the fluid. Fluid in the lungs can lead to aspiration pneumonia, which increases the risk for death.

A doctor may order a swallow study in which the person drinks several different types of liquids at different consistencies. The technician watches them swallow the fluids to see how thick the liquid must be to prevent the liquid from entering the lungs. Based on this test the doctor will be presented with recommendations for one of the following orders:

(1) Any consistency liquid
(2) Syrup/Nectar consistency thickened liquids
(3) Honey consistency thickened liquids
(4) Pudding consistency thickened liquids

Once the order is received all liquids that enter the person's body must be at the ordered consistency. If the doctor orders Syrup/Nectar consistency thickened liquids, the water or whatever they are drinking must be similar to the consistency of someone pouring pancake syrup out of a bottle. Honey consistency would be equivalent to the consistency of honey being poured from a bottle. Pudding would be similar to what pudding looks like and must be consumed from a spoon.

The product used to thicken liquids is called Thick-It. It is a powder similar to the consistency of baby formula. This product must be added to every liquid the person drinks or eats. Examples include: water, milk, juice, coffee, tea, soda, etc. All liquids must be thickened appropriately, according to the physician's order. In regards to food, all soups, gravies, etc must be thickened to the consistency ordered.

Tips:

- Allow 3-5 minutes for liquid to reach the desired consistency before serving.
- Thickened liquids continue to thicken while standing.
- <u>Never add ice to drinks that are being thickened.</u> Ice melts and creates water pockets.
- If thickened liquid is too thick, add additional liquid until the consistency is achieved.
- If liquid is too thin, add additional amounts of Thick-it until the consistency is achieved.
- Liquids may be thickened in advance and placed in a refrigerator to stay cold, however, you must check the consistency before the individual drinks the liquid to verify that it is the correct consistency.
- Check the consistency of drinks that have been sitting for any length of time in order to verify that the consistency is still correct.

Mechanically Altered Diet

This diet pertains to foods that are mechanically altered - such as blended, chopped, ground, or mashed. This makes the food easier to chew and swallow. Many people without teeth are prescribed this type of diet. Most of the Hamburger Helper meals meet this requirement as long as the pastas are cooked until soft, and the hamburger is in small pieces.

Food Group	Daily Servings	Food Allowed	Foods Not Allowed
Dairy	2-3 Servings	Milk products, yogurt, cottage cheese, soft cheeses, cheese sauce	Hard cheeses
Desserts and Sweets	Moderation	Smooth custards, pudding, sherbet, shakes, jelly, honey, sugar, chocolate syrup	Coconut, seeds, nuts, whole or dried fruits, fried, tough or chewy items, granola bars, pies, chewy candy, hard crunchy cookies, licorice, taffy, caramel
Fats	Moderation	Butter or margarine, cream, cooking fats and oils, smooth gravies and sauces, salad dressing, mayonnaise	
Fruit	2-4 Servings	Cooked or mashed ripe fruits without seeds or skins, mashed bananas, cantaloupe, fruit cocktail, canned peaches/pears, fruit juices	Citrus fruits, blueberries, cherries, grapes, pineapple, apples, dates, figs, dried prunes, raisins
Grains	6-11 servings	Soft breads, pancakes, muffins without seeds, waffles, stuffing, breads and crackers softened in soup, mashed potatoes, rice, noodles, soft pastas	Crackers, popcorn, chow mein noodles, taco shells, cake and breads with nuts, seeds, or raisins; pita bread, rye or pumpernickel bread with seeds, bagels, French or sour dough breads
Meat	2-3 Servings	Soft, cooked meat and poultry with gravy or sauces to help moisten, soft, flaked fish without bones, casseroles with ground meat, legumes mashed and moist, tofu, smooth peanut butter, scrambled eggs	Dry or tough cuts of meat or poultry, fried fish, fish with bones, hot dogs, sausage, bratwurst, pork chops, steak, crunchy peanut butter
Soups		Broth, bouillon, blended strained soups, cream soups	
Vegetables	3-5 Servings	Well cooked, soft vegetables without skin or seeds, mashed carrots, cooked beets, finely chopped, cooked greens, mashed squash, vegetable juices	Hash browns, fried potatoes, potato skins, French fries **NO RAW VEGETABLES**
Miscellaneous	Us As Desired	Ground seasonings and spices, tomato paste, mustard, ketchup	Nuts, coconut, popcorn, potato chips

Resource: Nutrition Therapy for Mechanically Altered Diet

Pureed Diet:

Pureed diets are prepared so that they are easy to swallow. Foods must be the consistency of pudding or moist, mashed potatoes that are free of lumps. When pureeing food, use butter, milk, gravies, sauces, sour cream, yogurt, fruit juices, or vegetable juices. <u>Avoid just adding water, this makes the diet bland</u>. Foods need to be separated on the individual's plate instead of blended together. Potatoes should be separate, vegetables separate, meat separate, etc.

Food Group	Daily Servings	Food Allowed (blended)	Foods Not Allowed
Dairy	2-3 Servings	Milk products, yogurt, cottage cheese, soft cheeses, cheese sauce	Milk or yogurt foods made with nuts or seeds, hard cheeses
Desserts and Sweets	Moderation	Smooth custards, pudding, sherbet, shakes, jelly, honey, sugar, chocolate syrup	Baked goods made with nuts, seeds or coconut; chocolate, butterscotch or peanut butter chips, marmalade
Fats	Moderation	Butter or margarine, cream, cooking fats and oils, smooth gravies and sauces, salad dressing, mayonnaise	
Fruit	2-4 Servings	All cooked or canned fruits without skins or seeds; fresh peeled apples, apricots, bananas, melons, peaches, pears, fruit juices, nectars	Fruits with seeds, membranes, or tough skins such as strawberries, raspberries, pineapple, orange and grapefruit sections, cherries, seeded grapes, dried fruits
Grains	6-11 servings	Breads and crackers without nuts, seeds or dried fruits; cooked or ready to eat cereals, pancakes, waffles, French toast, rice, pasta	Coarse, whole grain breads or breads with nuts, seeds, or dried fruit; granola, fried rice, fried noodles, potato skins
Meat	2-3 Servings	Cooked, tender meat, fish and poultry, infant strained meats, cooked legumes, tofu, eggs	Fried meats, sausages or other products with tough skins, poultry skin, fish with bones, anchovies, fried eggs, nuts, crunchy peanut butter
Soups		Broth, bouillon; all blended strained stock-based or cream soups	
Vegetables	3-5 Servings	Vegetable juices, well cooked or canned vegetables, cooked and peeled potatoes	raw or fried vegetables, vegetables with seeds, membranes, or tough skins, such as corn; celery or tomatoes
Miscellaneous	Us As Desired	Ground seasonings and spices, tomato paste, mustard, ketchup	Nuts, coconut, popcorn, potato chips, relishes

Resource: Nutrition Therapy for Mechanically Altered Diet

Medication Storage

Controlled Substances

Medication Disposal

Medication Errors

Medication Storage

Keeping medications locked in a cabinet, cart, closet or medication room is the safest method of medication storage. Your organization may not require that medications be locked under certain circumstances. Always follow your agency's policy and procedure regarding medication storage. The following are general guidelines regarding whether medications need to be locked or may be unlocked.

- Group homes must have medications locked
- ISL's (Independent Supported Living) homes
    - If the agency has a policy stating medications must be locked, they must be locked
    - If one person in the home is unsafe to be around unlocked medications, all medications in that home must be locked
    - If all Individuals supported in an ISL are safe to be around unlocked medications (as determined by the treatment team), the medications may be unlocked **unless** the agency has a policy stating that all medications must be locked
- All Schedule Two medications must be under double lock and key. An example of this would include a lock box inside a locked medication cart, closet or medication room.

Additional Rules about medication storage

- Stored away from direct light sources (direct sunlight or under a light)
- Kept away from heat sources (near a stove, furnace or in a garage)
- Kept away from high humidity areas (near a sink, in the bathroom)
- Stored away from cleaning supplies or chemicals
- Oral medications and non-oral medications must be stored separately. There must be something visual that separates oral medications from non-oral medications:
    - Zip lock bag
    - Separate storage container
    - Separate shelf
- If a medication needs to be refrigerated and medications must be locked for the home, the medications needs to be in an approved locked refrigerator designated for medications only, OR placed in a locked box and placed in the home's refrigerator.

Controlled Substances

Scheduled or controlled substances are medications that are restricted due to their potential for misappropriation. Schedule I medications are illegal and are at the highest risk for being misused. These include cocaine, LSD and PCP. Schedule II medications are generally high potency pain medications such as Percocet, Vicodin and Morphine. As the number increases, the potential risk of misuse decreases.

All controlled substances are regulated and must be carefully tracked. Follow your agency's policy for securing and counting these medications.

Not all controlled medications are narcotics. Due to Methamphetamine and the "war on drugs", some over-the-counter products are considered controlled substances (such as certain cough medications), even though the product contains no narcotic components.

Counting controlled substances varies according to the state. Become familiar with your state's rules and regulations. Most agencies count all controlled substances whether a schedule IV medication or a schedule II medication. All schedule II medications must be counted on a routine basis regardless of the state the individual resides in. Controlled substances should be counted every time the staff person responsible for these medications leaves and another person takes over the responsibility. At a minimum, the medications need to be counted at least once every 24 hours.

Rules in Regards to Counting Controlled Substances

- There needs to be two people present when counting controlled substances, the person starting their shift (taking over responsibility) and the person leaving.
- Both people need to visually look at the medications to verify that every pill is accounted for.
- Both people need to look at the medication administration record (MAR) to verify that controlled substances were that were reported as administered are accurately documented on the MAR.
- Both people need to sign the controlled substance log. Never sign the log indicating the count is correct unless you have physically counted the pills and everything is correct. Once you sign the log you are taking responsibility for the current medication count.
- If the count is not correct, look at the MAR and see if any doses of the medication was administered that was not documented on the MAR. If you cannot determine why the count is off, you must notify your supervisor.

Controlled Substance Record

Name of Individual supported _____John Doe_____

Medication _____Lorazepam_____ Dose __1 mg_____

Date	Time	# Dispensed Since Last Count	# At This Time	Signature	Signature
10-14-11	7:00am	0	30	Robert Palmer	Frank Dudley
10-14-11	3:00pm	2	28	Frank Dudley	Madison Smore

Medication Disposal

The following are examples of when medications need to be destroyed or returned to the pharmacy:

- Medication was dropped on the floor or contaminated in some manner
- Expired medications
- Physician discontinued a medication
- Prescription Changed

Never flush medications down a toilet unless the patient education information that comes with the medication instructs you to destroy the drug in this manner. Flushing medications increase the risk of it getting into the general water supply.

Federal Guidelines on destroying medications:

- Remove the medication from the container (prescription bottle or bubble pack) and place the medication in an impermeable, non-descript container such as a water bottle with the label removed.
- Remove the label on the prescription bottle or bubble pack and destroy the label. If the label is intact and later found by someone outside the agency, this could be a HIPPA privacy violation. The label should be cut into multiple pieces or placed in a confidential recycling bin approved by your agency.
- The medications that were placed in the non-descript container needs to have an undesirable substance added to it. This could be kitty litter, coffee grounds or sand. You can go one step further and even add water to a water bottle to decrease the chance of the medications being misappropriated. The container then needs to be placed in the trash to be taken to the land fill. Always place the container in the trash located outside of the ISL or Group Home.
- Take advantage of community pharmaceutical take-back programs which allow the general public to bring unused medications to central locations for proper disposal.

Medications that are destroyed need to be recorded on a medication destruction log. At some point you may be required to prove where every dose of a prescription medication has been documented as given or destroyed. The medication administration record will serve as proof for the medications administered and the destruction log will serve as proof for the medications destroyed.

Example: You received 30 tablets of Percocet for an Individual supported following a dental procedure. Three years later during an audit you are asked to provide accountability for all 30 pills. The medication administration record shows that 10 doses were administered. You would also have to produce the medication destruction log showing that 20 pills were destroyed. Without this documentation it may appear that the medication was misappropriated.

Medication Destruction Log

Individual supported _____Johhny West_____ Date _____10-13-11_____

Medication	Dose	Prescription Number	Quantity Destroyed	Signature	Signature
Percocet	5/325	C123456	20	Frank Smith, RN	Jane Joe

There needs to always be two people involved in destroying medications. One of the person's involved needs to be a nurse, pharmacist or state auditor plus a witness. The witness can be anyone over the age of 18 and in some way associated with the agency.

- Nurse + Witness
- Pharmacist + Witness
- State Auditor + Witness

Medication Errors

If you make a medication error, follow your agency's procedure for reporting and documenting the error. This will include notifying the supervisor, the nurse and the doctor. An incident report must be filled out for every medication error.

Types of medications errors could include:

- Failure to administer a medication. This could be because the person forgot or the medication was not available in the home (not reordered correctly).
- Giving the wrong medication.
- Giving the medication at the wrong time.
- Giving the medication to the wrong Individual.
- Giving the wrong dose of the medication.
- Giving the medication by the wrong route.
- Giving the wrong form of the medication (pills instead of liquid).
- Giving a medication without an order.

> **Medications errors may result in adverse reactions. These reactions could range from a minor rash to death**

Medication refusals are not medication errors. Individuals have the right to refuse medications. Always document the refusal on the medication administration record along with detailing attempts made in the daily log.

Methods to Prevent Medication Errors

1. Take your time when preparing medications. Turn off the television, do not talk on the phone or try to do any other task outside of setting up and administering the medications.
2. When you remove medications from a bubble pack, make sure the pill went into the medication cup. The foil on the back of medication cards often stick to the pill causing it not to go into the cup.
3. If more than one "bubble" from a medication card has to "popped", make sure the card is marked in some fashion to make the person more aware of this. Highlight the "2 tabs" on the directions or take a 2 x 5 card and staple it to the bubble pack indicating that 2 tabs need to be administered. Also indicate this on the medication administration record by underlining "2 tabs" or writing in large print "2 tabs".
4. If you have a hard time remembering to pass medications, set a timer in the home. This could be an alarm on a cell phone, microwave timer, stove timer or an inexpensive alarm clock specifically for the home.
5. Stay with the consumer and watch them take the medication. Make sure the medication has been swallowed or applied before walking away. If you are unsure if the medication was swallowed, perform a mouth check. You may be able to have the individual eat a cracker and drink additional fluid depending on the medication.
6. Only prepare one person's medication at a time. Set the medication up, administer the medication, document that the medication was administered, then move on to the next person. You are not allowed to set up more than one person's medication at a time.
7. Check the medication administration record to the prescription bottle or medication card three (3) times. This allows you to verify that you have the right consumer, right dose, right med, right route and that they are not allergic to the medication. All three checks must be done before the person receives the medication.
8. Never put your initials on the MAR indicating that the medication was administered until after the medication has actually been administered. If you initialed each box as you set up the medications, but were called away on an emergency, you have falsely documented that a person was administered medications that they did not receive.
9. The last row on a medication card is marked letting you know that you need to reorder the medication. When you "pop" the first pill in the last row of a medication card, you need to verify that there is another card of that medication available. If there is not, you must go through the reordering procedure according to your agency or notify your supervisor. If you are administering from a prescription bottle, this process should occur when there is 6 pills remaining in the bottle.
10. When picking up medications from a pharmacy, take a list of the medications that you will be picking up. Before you leave the pharmacy make sure you have all of the medications you were sent for.

Practice Session

1. True or False. Schedule II medications only need to be locked if one or more individuals in the home are unsafe to be around unlocked medications.

2. Which is the safest place to store medications?

    a. In a locked cabinet
    b. Under the sink
    c. In the garage
    d. With the cleaning supplies on an elevated shelf out of the reach of the individuals supported

3. True or False. If a medication must be refrigerated and the individual supported is unsafe to be around unlocked medications, the refrigerator must be locked.

4. True or False. All discontinued or expired medications need to be removed from the ISL or Group Home by flushing them down the toilet.

5. In the event of a medication error, which of the following should not occur:
    a. Contacting a nurse
    b. Contacting poison control
    c. Monitoring the person and only reporting the error if the person shows any signs of an adverse reaction
    d. Contacting your supervisor
    e. Completing an incident report

Guidelines For Observing And Reporting

Guidelines for Observing and Reporting

Any person providing services to individuals supported have a responsibility to report signs and symptoms observed while working with the person. The following signs and symptoms require further assessment and should be reported to a supervisor and nurse.

General Body
* Dehydration
* Chills
* Loss of appetite
* Allergic reactions
* Increased thirst
* Eating problems
* Weight gain/loss without effort
* Dizziness, weakness, fainting
* Severe headache
* Swelling
* Abnormal posture, gait, or movement

Vital Sign
* Elevated temperature
* Weak pulse
* Resting heart rate over 100
* Resting heart rate under 60
* Irregular pulse
* Wheezing/rattling respirations
* Difficulty breathing
* Shallow Breathing
* Deep respirations
* Painful breathing

Skin
* Any burns
* Dry/peeling skin
* Yellowing of the skin
* Rashes/hives
* Moles or sores with irregular borders
* Unusual bruising
* Wounds not healing

Eyes
* Drainage
* Bleeding
* Redness / irritation
* Loss of sight / decrease in vision
* Dilated or contracted pupils
* Foreign body in the eye
* Yellowing of the white part of the eye

Nose
* Difficulty breathing
* Nose bleeds
* Chronic discharge

Ears
* Loss of hearing
* Discharge from the ears
* Pain in the ear
* Excess ear wax
* Foreign body in the ear

Guidelines for Observing and Reporting (Continued)

Mouth / Throat
* Redness or white patches
* Swelling
* Missing teeth, chipped teeth
* Dentures not fitting correctly
* Toothache
* Difficulty swallowing
* Sores in the mouth
* Coughing after drinking / eating

Reproductive
* Drainage
* Foul odors
* Skin breakdown
* Complaints of painful urination
* Blood in the urine
* Itching, redness, swelling

Feet
* Cuts, sores not healing
* Thick, discolored toenails
* Ingrown toenails   * Swelling

Digestive
* Nausea or vomiting
* Distended abdomen
* Constipation or diarrhea
* Hemorrhoids
* Rectal bleeding
* Abdominal pain

Chest
* Pain
* Shortness of breath
* Chronic cough
* Difficulty breathing

Urinary
* Painful urination
* Inability to urinate
* Blood in urine
* Urine dark or cloudy
* Incontinence

This list is <u>not</u> all-inclusive.
Anything you see that you feel is not normal for the individual supported needs to be reported to a supervisor, nurse, or doctor.

Emergency Conditions Requiring an ER visit

- Poisoning (If directed by poison control)
- Uncontrolled bleeding
- Unconsciousness
- Cardiac arrest
- Respiratory failure
- Unable to breath correctly
- Head injury
- Fractures of long bones of arm or legs
- Uncontrolled seizures lasting past five minutes or as detailed in seizure protocol
- First known seizure
- Hypoglycemia not responding to food or interventions
- Severe allergic reactions (anaphylaxis)
- Person threatening suicide, with means of carrying it out
- Threats of harm to others, with means of acting it out
- Severe unexplained pain
- Hyperglycemia above 350 or outlined in diabetic protocol by the physician
- Person appears very ill, skin blue or gray
- Extremely low or high vital signs, abnormal for that individual
- Severe wheezing
- Bloody or coffee ground looking vomit or stool
- Broken bone with bone protruding through the skin
- Rapid onset confusion or disorientation
- Burns involving blisters or with skin coming off
- Fever greater than 101 F.

This list is not all-inclusive.

Call Poison Control anytime an individual ingests a toxic substance (ie: cleaning supply, gasoline) or ingests the wrong dose or wrong person's medication when there is a potential for poisoning

Medication Aide Training for Non-Licensed Personnel

Contract

Outlined in this contract are the expectations in regards to passing medications to individuals supported.

(1) Only people who have successfully completed the Medication Aide training for Non-Licensed Personnel and have maintained the required two year updates may administer medications or supervise self administration of medications.
(2) All aspects of the Medication Aide Training Program will be followed.

In accordance with the Medication Aide training for non-licensed personnel, you acknowledge that you have received training and will adhere to the following:

a. Only set up and administer one person's medication at a time.
b. Check the medication container and the M.A.R. three (3) times before administering any type of medication. Compare the name of the person to the medication, verify the dose to be given, the route, the time, the form of the medication (pill, liquid, etc) and that it is the correct prescribed medication. Also verify that the person is not allergic to the medication and that the medication is not expired.
c. Verify that you are administering the medication to the correct consumer; either by asking the consumer their name, having another staff person identify the consumer for you, or by other methods approved by your agency.
d. Observe the consumer taking the medication. Stay with the Individual until all medications have been taken; verify that all medications have been swallowed or properly administered.
e. Medications will be given on time according to the agency's policy and procedures.
f. When setting up medications, you verify that the medication went into the medication cup and is not stuck to the packaging or fallen on the floor.
g. Never sign the M.A.R. until after the medication(s) have been administered.
h. When using medications from the last row of a medication card, verification of additional medication availability will be made. Notify your supervisor, if necessary, to make certain that the medication will be available for the consumer as prescribed.
i. Never administer medication you are not familiar with. You must know the indication for the medication and possible side effects to observe for.
j. If something does not look correct with the medication or directions on the medication label, you agree to verify the directions or identify the medication with a nurse or a pharmacist before administering the medication.

Passing medications is a great responsibility. Adhering to the principles taught in the medication class is extremely important for the health and safety of the Individuals supported. By signing below you acknowledge that you have read this contract and are agreeing to pass medications in accordance to the class, and will adhere to everything listed in this contract.

Employee (Print) _____ (Sign) _____ Date _____

Instructor (signature) __ _____ Date _____

Testing and Database Storage of Certificates

- Agencies may order tests or create their own test for students who have completed this course.
- Instructors should monitor all students taking the test.
- Certificates may be issued by the agency **or** copies of the tests can be sent to Palmer's Publishing Company where a database of all certificates is maintained that have been issued and turned in.
- Certificates will be issued by Palmer's Publishing Company for all completed tests with a grade of 80% or above that have been mailed in.
- Certificates (at no additional cost) will only be issued if tests are ordered and returned through Palmer's Publishing (no photocopies of the tests are permitted if certificates are requested through Palmer's Publishing).

Address for ordering tests or returning tests for certificates:

Robert Palmer, RN, BSN

8060 South Turtle Creek Lane

Columbia, MO 65203

Please indicate the quantity of tests needed.

Please note that there are four (4) versions of the test available.

Indicate the number of versions of tests wanted (Example: If you order 12 tests do you want all tests to be the same version or do you want 3 of each version).

Cost of tests:	Quantity	Price
	Quantity 1-5	2.00 per test (Includes shipping and handling)
	Quantity 6-10	1.50 per test (Includes shipping and handling)
	Quantity 11-20	1.25 per test (includes shipping and handling)
	Quantity 21+	1.00 per test (includes shipping and handling)

Works Cited

American Diabetes Association, Diabetes Pro, Professional Resources On-line. http://professsional.diabetes.org

A Practical Guide to Clinical medicine, University of California, San Diego. 2011.

American Red Cross, First Aid/CPR/AED Participants Manual. Staywell Health & Safety Solutions. 2011.

Better Medicine, Psychiatric Disorders. Medical Reviewer: Brian McDonough. 2011.

Centers for Disease Control. http://www.cdc.gov

Epilepsy and Seizure Information for Patients and Health Care Professionals. www.epilepsy.com

Nurse Assistant in a Long-Term Care Facility, Student Reference. 2001 Revision. Instructional Materials Laboratory. University of Missouri-Columbia.

Nursing 2011 Drug Handbook. Lippincott Williams & Wilkins. 31st edition.

www.ingramcontent.com/pod-product-compliance
Lightning Source LLC
Chambersburg PA
CBHW080918170526
45158CB00008B/2159